Space-Time
Continuum

by *Frank A Santora*

Edited: Cheryl L Deutsch

FrankS34@ptd.net

Copyright©2019 All Rights Reserved
published *Amazon.com* bookstore

Table of Contents

Preface 3

Introduction *"Welcome to the strange world of Relativity"* 4

Chapter 1 *"Space-Time Continuum ...What is it?"* 7
- *Visualizing Space-Time Continuum* 7
- *4-D Space-Time Beginnings* 9
- *Enter: Space-Time Continuum* 11
- *ALERT! ⚑ Flaw #1, "drip, drip, drip"* 16
- *Additional remarks* 20

Chapter 2 *"Run! The foundations are crumbling"* 22
- *Postulations 1 & 2* 22
- *Reference 1: A must reading* 26

Chapter 3 *"ALERT! Special Relativity flaws ahead?"* 28
- *ALERT! ⚑ Flaw #2, "Appearances are deceiving"* 28
- *Chasing a light beam* 29
- *ALERT! ⚑ Flaw #3, "Contraction contradicted"* 31
- *Enter time dilation* 33
- *ALERT! ⚑ Flaw #4, "SR singularity singularly wrong"* 34

Chapter 4 *"ALERT! General Relativity flaws ahead"* 35
- *The Equivalence Principle* 36
- *ALERT! ⚑ Flaw #5, "Deceiving bouncing balls"* 37
- *Additional Concern* 38
- *ALERT! ⚑ Flaw #6, "False-Negatives=Proof-Positive"* 38

Chapter 5 *"Oh, the error of our ways"* 40
- *Opposing Relativists* 40
- *Advance of the perihelion of Mercury* 45
- *CAUTION! ⚑, "Computer age ahead"* 47
- *Deflection of light by a gravitational field* 48
- *CAUTION! ⚑, "Misgivings around the corner"* 49
- *Gravitational RedSshift* 52
- *CAUTION! ⚑, "That shifty Red Shift"* 53
- *Gravity Probe-B (GP-B)* 55
- *Error sources* 59
- *CAUTION! ⚑, "The rest of the story"* 60
- *Statistics story* 62
- *CAUTION! ⚑, "Statistical False-Positives"* 65

Table of Contents, continued

[Chapter 6](#) **"The twin paradox: What is it?"** 70
- *Riding A Light Beam* 70
- *ALERT!* 🚩 *Flaw #7, "Don't blink, you'll miss a flaw"* 72
- *Passing A Light Beam* 73

[Chapter 7](#) **"The twin paradox debunked"** 75
- *All Aboard for round trip to Alpha Centauri* 75
- *Round Trip Objective* 78
- *All Aboard, ready for departure* 80
- *Voyage Cruise Details* 80
- *CAUTION!* 🚩 *Com #1: At the outbound half-way mark* 83
- *CAUTION!* 🚩 *Com #2: At entry into Proxima-b orbit* 85
- *CAUTION!* 🚩 *Com #3: At the inbound half-way mark* 87
- *ALERT!* 🚩🚩 *Flaw #8, "Twin paradox debunked!"* 88
- *Paradox Debunked!* 93

[Summary](#) 94

Preface

It is without question that this book will elicit a lot of commentary, both good and bad, from experts and interested people as well. Herein, Einstein's various "thought experiments", that led to the concept of **Space-Time Continuum**, are more thoroughly examined and extended. Upon closer examination, a number of very serious flaws were uncovered in the theories of **Special Relativity (SR)** and **General Relativity (GR)**. As a consequence, the combined flaws essentially lead to serious "cracks" in the overall Relativity theory and the subsequent "***shattering***" of the **Space-Time Continuum** concept. This book does not present complicated and boring physics or mathematics. Instead, it approaches the subject the same way that Einstein did *before* he developed his mathematics; -–through visual "thought experiments". So, sit back and enjoy a return to reality.

Introduction

"Welcome to the strange world of Relativity"

When we say that Einstein's theory is non-intuitive and hard to digest, we are not kidding. On June 30th, 1905 he published his original paper entitled: *"On the Electrodynamics of Moving Bodies"*. In his paper, he introduced his first postulation (assumption) on the principle of **Special Relativity (SR)**; that *the measured speed of light (c) is constant to all observers whether the observer is "fixed" inertially or moving.* This immediately caused a controversial stir in the scientific community. A second paper, presented in 1907, extended **SR** to include a **General Theory of Relativity (GR)** Suddenly, the then current world of physics went into turmoil over pondering the implied results of his publications; and ever since, has been trying to prove his theory with various experiments and further mathematical extensions, interpretations and innovative concepts. One awe inspiring concept is that of a **"Space-Time Continuum"**. We'll get to this concept later in the Chapters below. Needless to say, to this day, controversy has been ongoing in the scientific community. There are many reputable scientists, engineers and organizations that either support or challenge the validity of **SR** and **GR**. However, the current consensus of the greater scientific community is that the theories of **Relativity (SR & GR)** have been proven valid.

Before we proceed, let me first give you a preliminary visual illustration of what this world of **Relativity** looks like. ***Illustration 1, "Welcome to the World of Relativity"*** below (modified from artist S. Dali) renders the world of **Relativity** as observed by a spectator on Earth looking at a space traveler. Need I say more? In the **SR** world there is a speed limit ***c** (speed of light)*, which cannot be exceeded; other wise, the person traveling in space would shrink to nothing. Further more, the traveler's on-board physical clocks slow down, and objects shorten in length and increase in mass; ...but the traveler in space perceives no such changes. Everything appears normal to him. Let's say that the space traveler has a twin brother back on Earth. When he returns to Earth, his twin brother will be

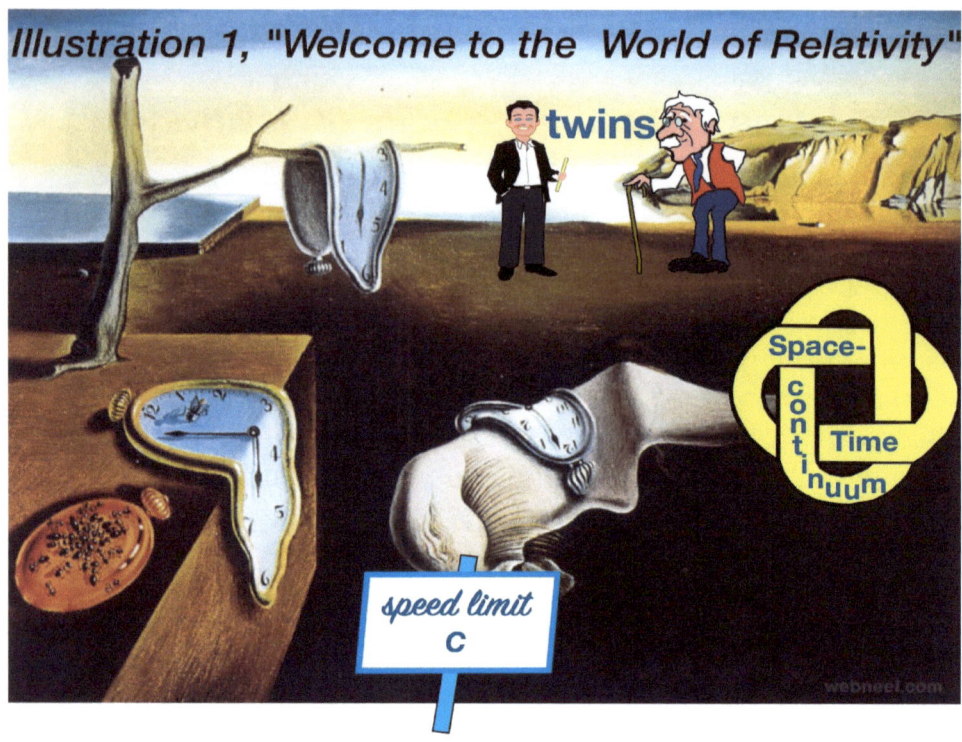

Illustration 1, "Welcome to the World of Relativity"

much older than himself. Actually, this relativistic "aging process" affects all the inhabitants on Earth. In addition, **GR** generated a four-dimensional definition **(X, Y, Z, t[time])** of a curved space that appears to be present as an inherent "mesh" throughout the Universe. This concept is referred as the "**Space-Time Continuum**" and has some mathematical rigor stemming from both **SR** and **GR**. It also has implications, especially when considering the relativistic "aging process", and possible methods to achieve greater travel speeds in space.

In the following Chapters, we will point out to the reader what this author considers to be significant "flaws" present in **SR** and **GR** thereby potentially invalidating these theories. I know this is a forward statement to make, but I will caveat it with an open mind to accept suggestions and realistic quantitative objections, but not any "off-the-cuff" Relativistic-position opinions.

The Chapters below will employ visual illustrations of various **SR** and **GR** "thought experiments" rather than fanciful, and elegant complicated mathematics and physics. Any mathematical expressions or physics principles will be minimized and used only to make a point of clarity or validation. If you abhor equations, just skip over them, they are not necessary to enjoy or understand this book.

Now, any reader from nine to ninety is invited on an exciting visual tour that he/she can easily understand. So be ready to enter an unbelievable weird world of paradoxes and controversy.

Chapter 1

"Space-Time Continuum ...What is it?"

What is **Space-Time Continuum**? It's easier to explain what it is not! It is not easy to visualize. It is not easy to comprehend. It is not intuitive, and it definitely is not tangible! But, let's give it a try anyway. It's a concept stemming from a motivation to combine space, time and gravity into a universal and unified physical model of our world. This concept is not new. It traces its beginnings back to the late 1800's and early 1900's. This time period is the period when the worlds of Isaac Newton and Galileo were extended to include the world of Relativity. The world of Relativity included notable physicists and mathematicians such as Hermann Minkowski, Hendrik Lorentz, Albert Einstein, Henri Poincare, Neils Bohr and others. These renowned physicists were all contemporaries during this time frame and were responsible for major advances in physics leading to advancements in electromagnetics, **theory of relativity**. and the concept of a **Space-Time Continuum**.

Since this book is challenging the foundations of this concept, we will first explore the concept to obtain a better understanding. The way to accomplish this is to dive right in! Here we go!

Visualizing Space-Time Continuum
A lot of authors have tried to present a visualization of what **Space-Time Continuum** looks like. The most popular depiction presented is shown in the following *Illustration 2, "Space-Time Continuum"*. Let's start by first looking at the web like background in the *Illustration 2*. This background represents a flat spatial two-dimensional (2-D) plane of X-Y coordinates. This plane is commonly referred as the "fabric" of space and can consist of any number of coordinate pairs as desired. Note that the depiction contains only one single plane. Now, if there is a sizable mass present, the fabric is depressed as if a weight is resting on the surface. This is why the fabric is also sometimes expressed as a "trampoline". For example, a mass can be the sun, a planet, or even a galaxy. The presence of mass illustrates the curvature

aspect of space; but the fabric remains essentially undisturbed in the more distant regions.

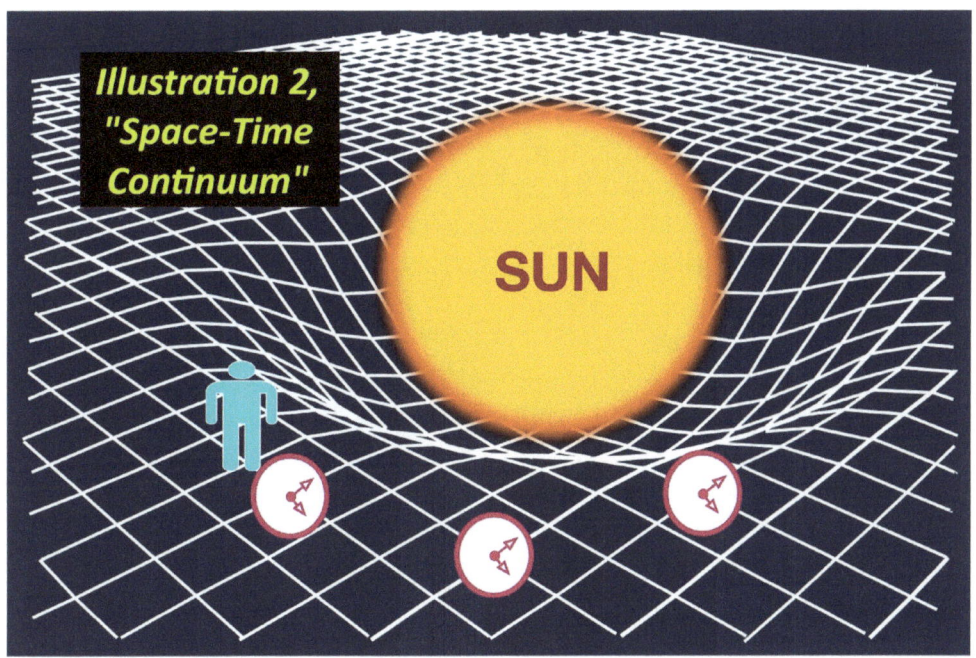

To gain some insight to the dynamics of Space-Time, imagine a man or space ship inertially at rest some distance away from the space curvature with a clock by his side. All the clocks on the plane read the same time. The man is standing perfectly still and the clock keeps advancing second by second. The plane he is standing on appears not to have changed; but...it has! The plane he was standing on is now somewhere "below" him — in the past. I told you it would be difficult to comprehend, but let's move on. Let the man or Starship begin to move toward the mass (in this case the sun) at some speed. When looking at the above illustration, you would naturally assume that he would begin to slide physically downward along the space curvature contour. Not so fast! You might ask, "Wouldn't he just walk or fly beneath the sun and arrive back on the same apparent plane?" This is non-intuitive since you may further ask, "Doesn't gravity pull him in and cause him to impact the sun?" Apparently, the illustration is insufficient in conveying the desired attributes of **Space-Time Continuum**. We must delve further into the concept to obtain a much clearer understanding. To accomplish this we will address, unfortunately, its mathematical beginnings.

4-D Space-Time Beginnings

In the beginning, Hermann Minkowski orally presented (1909) an essay titled "Raum Und Leit" which was first translated as "Space And Time" by Edward Carus (1918). An annotated version of this essay has recently been published (March 10, 2013) by "The Net Advance Of Physics", article: "Space and Time". The concept was advanced and is represented by a four variable equation as;

$$V^2 t^2 - x^2 - y^2 - z^2 = 1$$

where x, y, & z are the 3 geometric spatial variables and t is the 4th variable (time). V is not a variable and is a constant velocity for a specific case in question. One can see that the above equation is basically a hyperbola and defines travel lines and associated times in space. It is very important to note that the equation is abstract in nature, and is not observed physically anywhere as having real physical substance. Like geodetic lines of latitude and longitude superposed on the Earth's surface, they are mathematically imagined and ascribed for navigational purposes. We can now show the meaning of this equation in the following ***Illustration 3, "4-D Space-Time"***.

In *illustration 3*, V is assumed to equal the speed of light c (186,000 statute miles per second) since most physicists and folks interested in Relativity deal primarily with visual observations. It must be noted here that V is not limited in any manner. Its value could be $V=0$ to $V=\infty$ (infinity), and there is no singularity present. Since V can be in any direction, a cone is envisioned that contains the space *(x, y)* area that can be reached for V less than or equal to c. Likewise, for curiosity's sake, expansion of the coverage cone for $V=2c$ is indicated. The attainable *x-y* planar area is indicated by the horizontal "present" *(t=0)* Space Plane in the Illustration. The coverage cone below the Space Plane represents the past (time already passed); and the coverage cone above the future (time to yet occur). Time *(t)* always moves forward. One cannot go back in time. For example, a person standing at *x-y* coordinates 1 second prior to the present would have had coordinates of *(0,0,0,-1)*. Note that if he remains standing in the same place, his "life path" or "time-history" from past to the present is now *(0,0,0,0)*. This time history is fixed as part of the past and cannot be altered, even though scientists have pondered the possibility of traveling back in time.

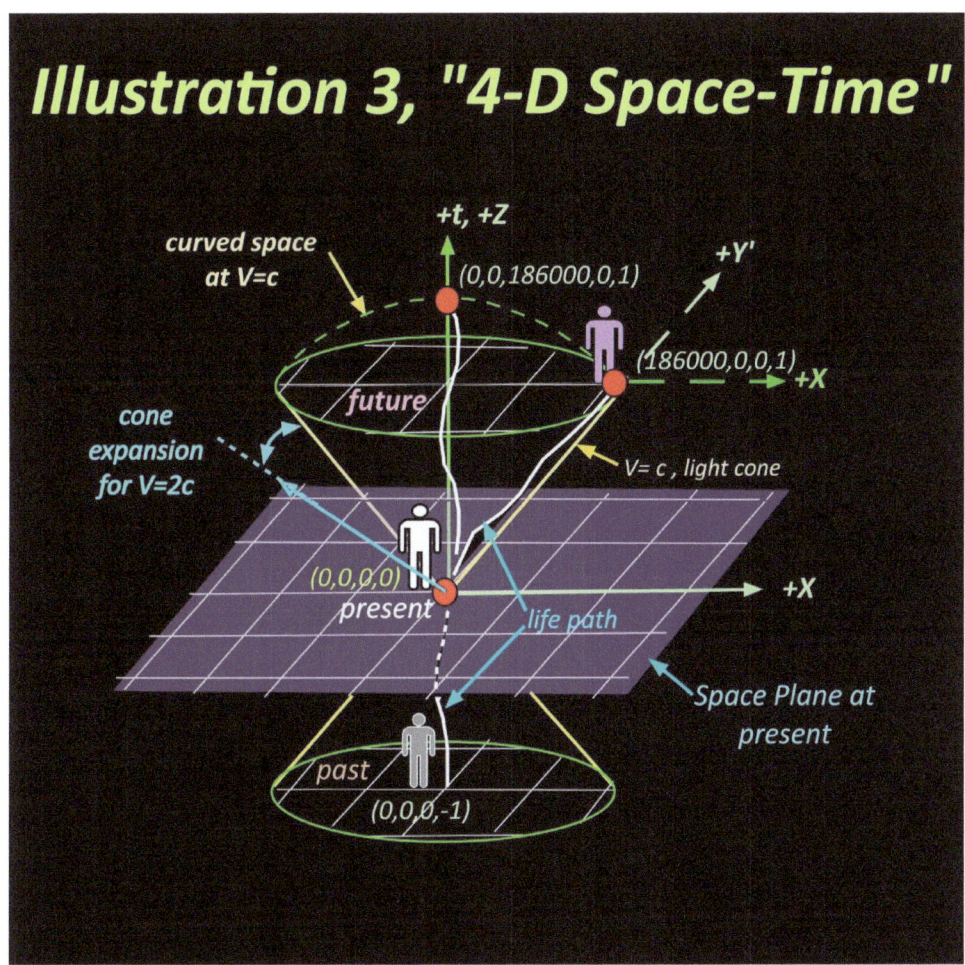

The light cone shown above the plane represents all possible locations that can be reached traveling at the speed of light or less. The future life path shown is only one of millions of paths that the person can potentially travel. If a person travels at speed *c* for one second *(t=1)* along *x*, then after one second, be careful watch this; in his new reference frame *(x', y' z', t')* the coordinates now become *(186000, 0, 0, 1)*. But the *(x', y' z', t')* can now revert to the "present" again *(x, y, z, t)* and a new current Space Plane is defined *(0,0,0,0)*. It is very important to understand that if the path involved a movement in the *z* direction, the Space Plane is spatially translated by the amount *z*, thereby resulting in *z'*=0, or again, *z*=0.

In engineering circles, the complex **Illustration 3** can be simply represented by cartesian **XYZ** coordinates with any path *x'y'z'* (past or future) with time *(t')* ticks along the path as shown in **Illustration 3A, "An Engineering view of 4-D Space-Time".** This illustration is included, without further comment, for the convenient comparison by the reader

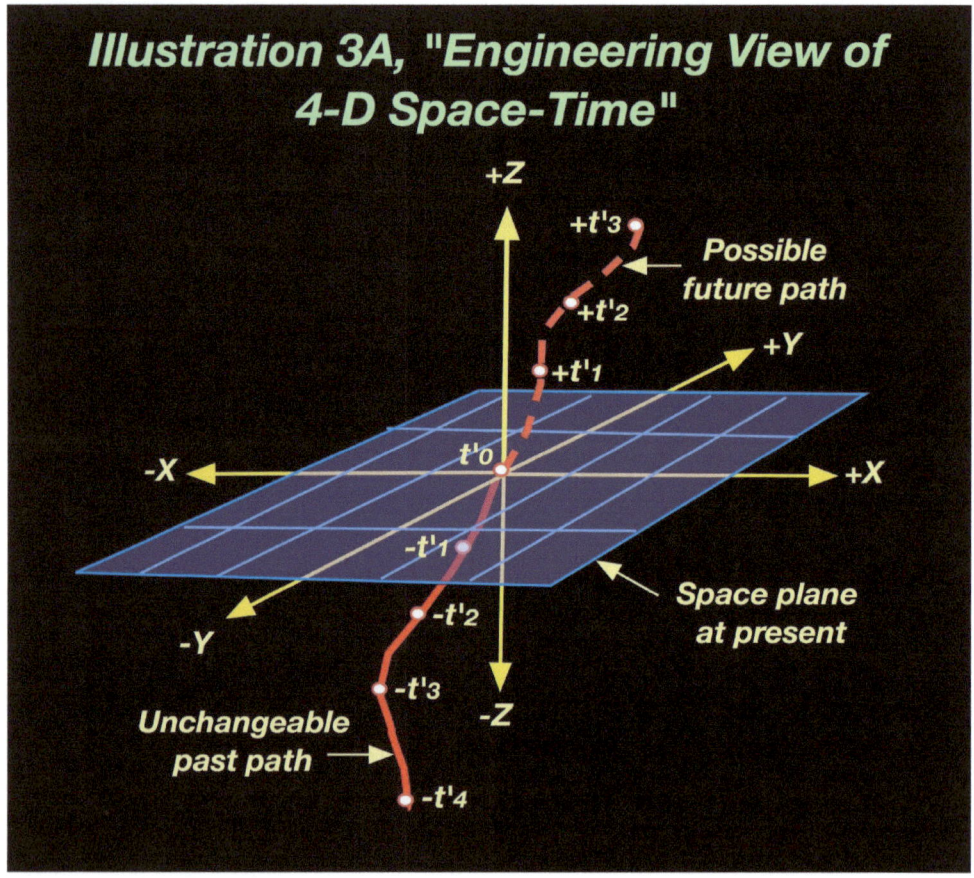

Now the big question arises, how do mathematical imaginary abstract lines turn into a "real mass" or "gravity" effects field??

<u>Enter: Space-Time Continuum</u>
Moving from the abstract to a theoretical real world takes a "really" strong imagination. But Einstein did exactly this by starting with a

vague idea and extending with factual physics, and brilliant mathematics. The result was the **Space-Time Continuum theory**.

One day, while looking out his apartment window, he watched a man working on the roof of a nearby building, He visualized the man falling toward the ground and wondered why the falling worker felt no force pushing him downward. He then questioned the current accepted reason of attraction between bodies namely; gravity. Somehow, he thought there might be a different explanation in that there was no known force to cause an acceleration of a falling body (from Newton's famous *F=ma*). Based on his previous postulation of **General Relativity (GR)** concerning the equivalence of acceleration *(a)* and gravity *(g)*, he began looking for the mysterious force *(F)* that eluded Newton and now, himself. His challenge was how to incorporate the "effects" of mass *(m)* into the abstract realm of Space-Time.

He began his quest, in all of dubious places, fluid mechanics. He applied the known principles of fluid mechanics known in that day; and which have been improved continually to this day. This Physics subject addresses all aspects of fluid flows including pressure, flow velocity, density, stress tensors, etc. The analytical fluid mechanics equations available today, e.g. Navier-Stokes, are extensively used in the Aeronautical and Maritime industries. Einstein was able to, right or wrong, relate and apply fluid mechanics to the abstracts mathematics of Space-Time thereby giving it a real physical reality. However, it is still a theory although it is overwhelmingly accepted.

This author includes one proven fluid mechanics equation used by Einstein as a starting point. This equation is shown below and is referred to as the "Bulk Modulus" *(K_b)* equation. It is given here not to mathematically prove anything, but only to show relationships to the physical world.

$$K_b = -Vol \cdot (\Delta P / \Delta Vol)$$

Vol represents a **closed** volume of an object, like a solid planet. *ΔP* is a change in pressure. *ΔVol* is a change in object volume.
This equation simply means that the displacement of Space-Time made by a closed volume *(Vol)* exerts a positive pressure *(ΔP)* on

the surface of the volume, which then results in a volume decrease *(-ΔVol)*. This "process" is shown in **Illustration 4** below.

In **Illustration 4, "Space-Time Curvature Pressure"**, the underlying development theme based on a closed volume is:

closed volume >> leads to
curved space-time >> leads to
surface pressure >> leads to
"g" field effects.

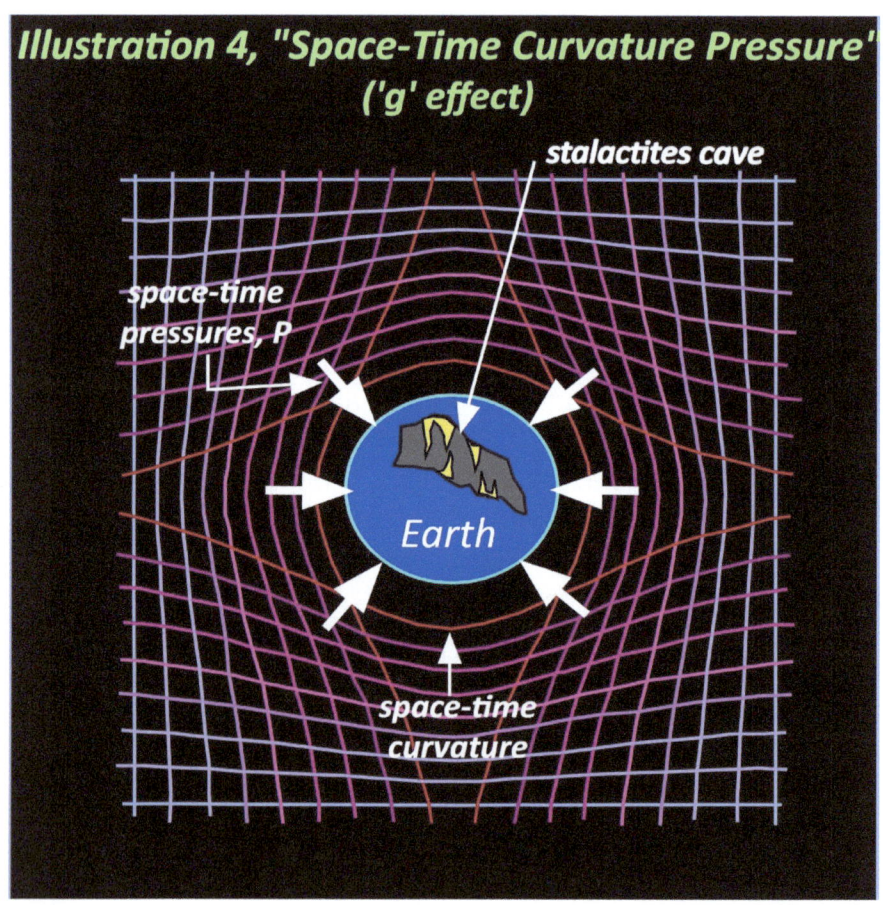

Pseudo *"g"* field effects are incorporated by the gradient exertion of pressure on the surface which can exhibit a corresponding mass *(m)* effect that has all the characteristics of *real **m***. This is not very obvious, but is mathematically sound. It should also be noted that in

SR, mass energy is intricately involved with the total momentum of an object. For the case of acceleration of objects in the **Continuum,** Einstein applied the equivalence principle (Postulation 1) to his field equations. His field equations utilized an energy-momentum mathematical tensor to connect the attributes off mass-energy to the **Continuum**. Any way, the result is that the **4-D Space-Time Continuum** could now be expressed as a representation of the entire universe, namely:

$$m=f(x, y, z, t)$$

So, it is obvious that both **SR** and **GR**, although not prominently apparent, play a foundational role in this final universal equation.

Notice in the Illustration that a cave with stalactites is included. We will return to this cave in a minute; but first let's look at a more exacting visualization of the **Space-Time Continuum** in the next *Illustration 5, "Multi-Plane Space Curvature"*.

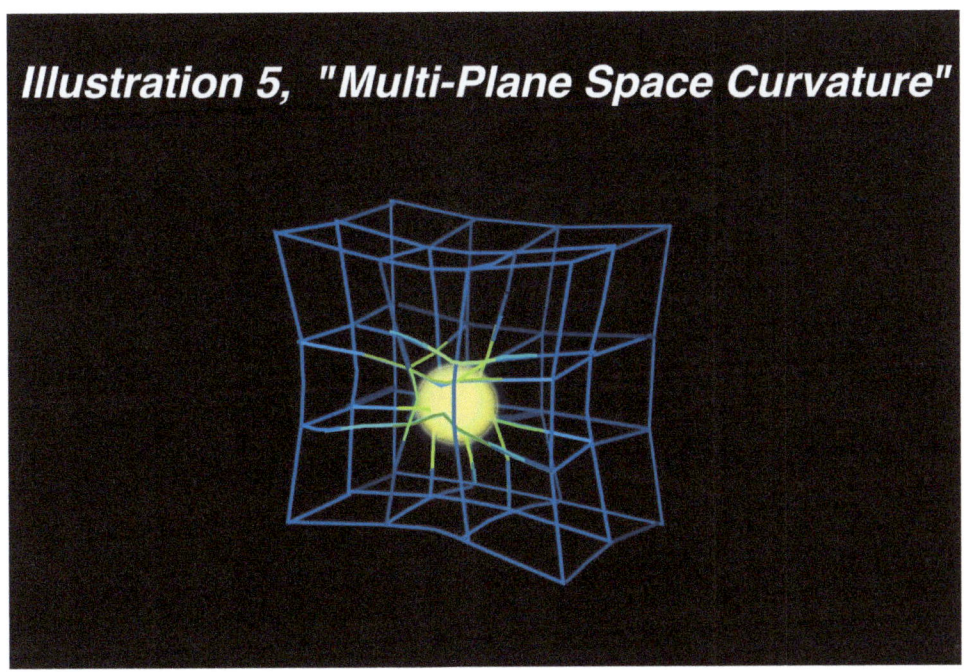

This Illustration portrays a more realistic three-dimensional view that helps visualization immensely. It shows several Space-Time planes both in the horizontal and vertical positions i.e., orthogonal to each other. The curvature effect is pronounced and is a result of a **closed** celestial body like a star. Hence, the connection of coordinates of all the planes now becomes a "mesh". This mesh is the full "fabric" of **Space-Time Continuum**. Note that the illustration is "static" and does not include other theoretical Space-Time "dynamic" effects. These include such effects as "geodesic curvature displacements" and "Space-Time frame dragging" due to spinning bodies (so earnestly sought by the Gravity Probe-B satellite experiment), and the compression of the mesh due to the orbital velocities and acceleration of the bodies in the **Continuum**.

What do real life satellite or small body (orbitals) trajectories look like in a **Space-Time Continuum** with *m* effects?". To get a feel of trajectory movement, you can refer to *Illustration 6* below. In *Illustration 6, "Trajectory Profiles"*, four space curvature planes are shown.

Illustration 6 , "Trajectory Profiles"

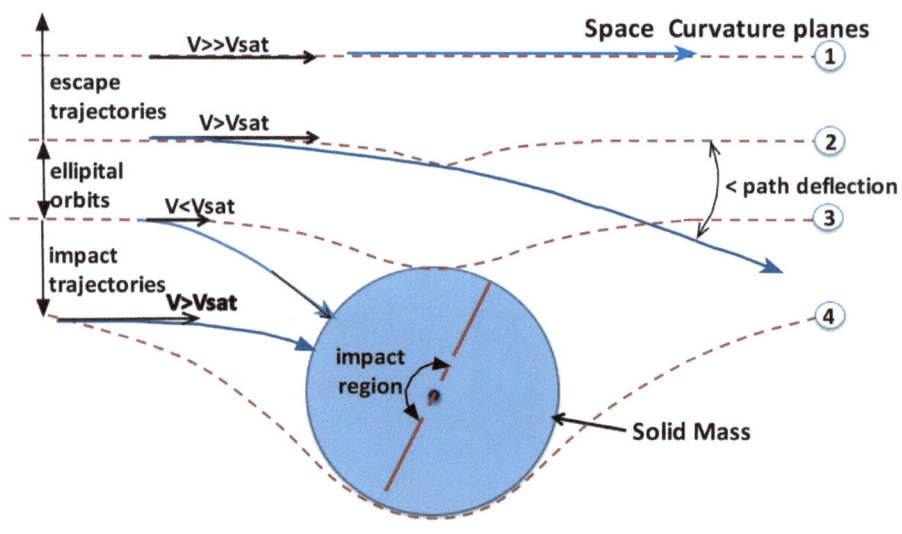

.
Plane 1 is a flat plane, without curvature, far from the disturbing mass (e.g. Earth) and therefore without any "gravity" effects. Any

trajectory in this plane will continue to move with a uniform non-accelerating velocity. Remember, space curvature is really a "pressure" field that exhibits a *"g"* field gradient effect.

Plane 2 is also remote from the Earth but close enough to have some curvature due to *"g"* effects. Trajectories in this plane will generally have velocities greater than satellite speeds and their paths will only be deflected.

Plane 3 has a more pronounced space curvature, and their trajectory paths may or may not impact the Earth depending on their initial speed. The region between planes 2 and 3 represent typical elliptical orbits with velocities between satellite and escape speed.

Plane 4 shows a curvature where the remote approach velocity *(V)*, if path were not deflected, would impact the Earth. In this case, any velocity between zero and infinity would be an impacting trajectory. Although the above Illustration cannot represent complete trajectory behavior, it nevertheless does provide ample insight.

It is now time to move on. In summary, what Einstein arrived at is that <u>the roof top worker did not fall due to the pull of gravity, but rather, by the space-time field pushing him downward</u>; thereby proving that Chicken Little was right after all (just joking). This vision is certainly not intuitive at all.

By now the above discussion is more than you've ever wanted to know about the **Space-Time Continuum**. It doesn't make much difference in the every day life of the general public, nor does the theory of Relativity. Nor does it make any difference to orbital mechanics engineers who design space missions. Practically speaking, **Space-Time Continuum** is merely an artifice of expressing *"g"* effects in a different manner. It makes trajectory analyses unacceptably complicated and awkward. Newton's definition of gravity (as the mutual attraction of bodies) is more than ample and effective in planning and designing space missions. But before we leave this discussion, there is one last very important observation to make; and that is, it is flawed! It doesn't make sense!

ALERT! 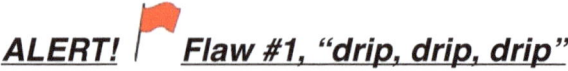 Flaw #1, "drip, drip, drip"

"Well why doesn't it make sense?"
"You have to listen carefully."
"Do you hear it?

..... Drip ......drip ......drip ....,

"Is it a leaky faucet?"
"No!"

Drip ....drip .....drip .....

"Is my roof leaking?"
"No!"
"I think it's from my basement."

Drip ....drip .....drip .....

"Well I'll be!" "It's coming from underneath the ground!"

Now it's time to go back to the **Illustration 4, "Space-Time Curvature Pressure"** and take a closer look at the cave located **inside** the Earth. **Illustration 7, "Stalagmites & Stalactites Cave"** shows a rendering of such a cave.

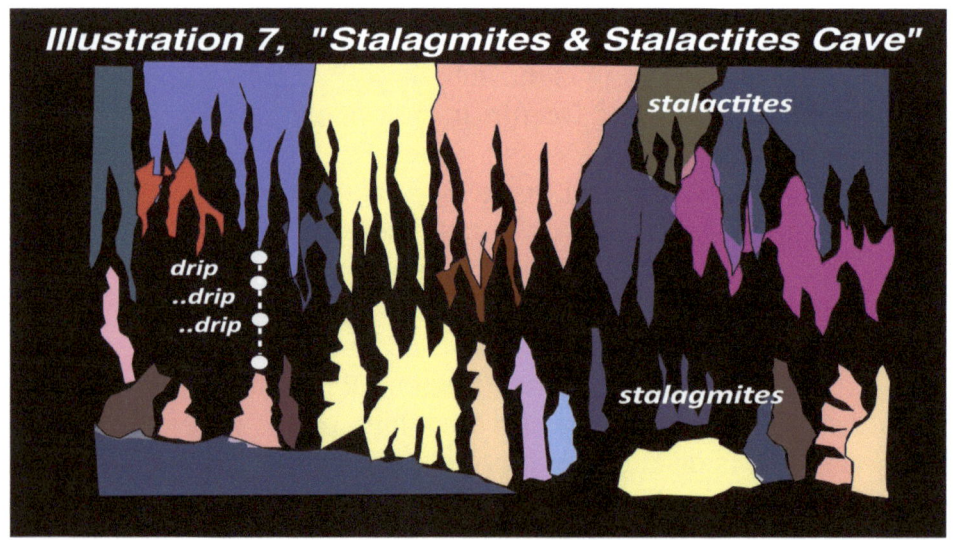

The first thing to notice is that there is a slow, steady drip, drip, drip from a stalactite on the *ceiling* of the cave. Water is dripping from there onto a stalagmite on the cave *floor*. This calcification process has been going on for thousands if not millions of years. You might ask "What's so important about that? There are many such caves in the earth." The importance is that it proves the concept of **Space-Time** theory to be **flawed**! "How so?" you might say. Well the slow dripping proves that gravity effects are present <u>within</u> the Earth.

Space-Time theory, as developed by Einstein, mathematically does not allow for gravity effects within the Earth. Let's go back to the ***Illustration 4, "Space-Time Curvature Pressure"***.

As theoretically and expressively stated, the cause of a **Space-Time curvature** is the presence of a **closed volume** (e.g. Earth). Earth is an example of a closed volume having a radius *(R)* of 3963 miles and a spherical volumetric displacement of **4/3πR³.**

In the referenced illustration, **Space-Time curvature** is "wrapped" around the Earth's surface ...<u>not</u> <u>below</u> the surface. Consequently, the resultant Space-Time 'g-field pressure' acts <u>only on the Earth's</u> <u>surface</u>. This pressure would then be transmitted mechanically to the the inside of the Earth and distributed via the solid mass within. Cavities or caves within the Earth <u>cannot</u> distribute any of these compressive stresses.

There are several more flawed constructs that can be made by applying the **Space-Time Continuum**. One of them, if it wasn't flawed, would have Einstein inventing the first anti-gravity shield. Construct a completely <u>empty</u> sphere strong enough to withstand any stresses from a space curvature field. Now just enter the steel sphere and close the door behind you. You should begin to float in a weightless environment, because any manifested space field is external to the sphere and physically transmitted only through the sphere's shell structure
.

A second construct involves a small exposed cave entrance thereby implying that entry of the space curvature field will occur. This is really a picayune nit. Average cave entrances are one hundred feet or less. This amounts to less than 0.1×10^{-6} % of the Earth's Surface for one thousand caves. If this is still a concern for any one, just seal the entrance with a steel bulkhead and verify that ***"g"*** effects are not present. What would you guess the results of these two constructs would be, and would you bet your life on it.

Along the same line of thinking, engineers have been considering building a high speed 3000-mile tunnel rail system from New York City (NYC) to Los Angeles (LA). The passenger gravity train would require very little energy to make the trip. For the most part, it would rely on the component of Earth's gravity acting along the rail track direction. Maintaining a partial vacuum inside the tunnel would reduce aerodynamic drag effects. ***Illustration 8, "NYC to LA Subterranean Gravity Train"*** depicts such a train beginning at NYC with a gravity component of about 0.4g that accelerates it toward LA. Drag effects would be offset via augmenting using rail-train electromagnetics similar to the Japanese high-speed trains. As the train moves down the tunnel, the accelerating gravity component reduces to 0.2g's and reaches zero acceleration as it crosses the trip half way mark. At this point, the train has acquired a very substantial speed and begins to slow down as the gravity component begins to act as a brake. Speed continually reduces to zero as the train reaches LA. Average speed of the trip would be 200mph to 350mph with a trip duration of 10 hours.

Illustration 8, "NYC to LA Subterranean Gravity Train"

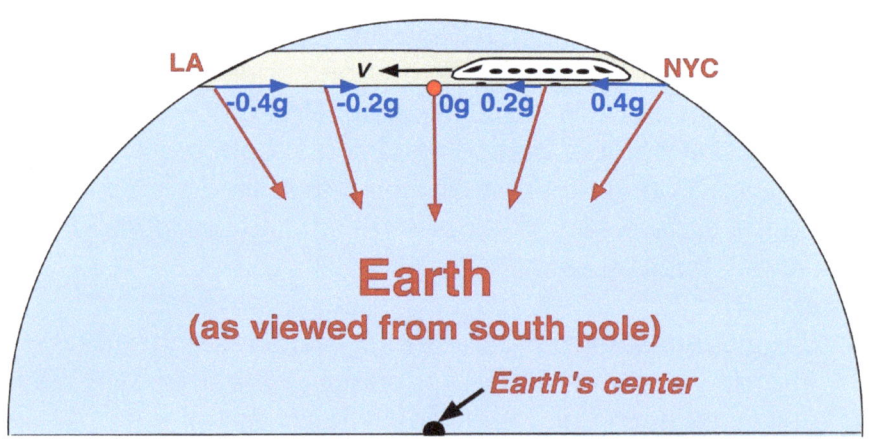

Now the whole process reverses in making the return trip to NYC. As mentioned above, additional propulsion will most likely be needed to counter air drag and any rolling friction. The point to make again is that gravity exists <u>inside</u> the Earth and is not the result of any pressure from the external space curvature.

It's concluded that a serious flaw exists with the **Space-Time Continuum theory** as currently espoused and developed. In the following chapters, additional flaws in the theories of **Relativity** will be presented that will further substantiate the incorrectness of all these theories.

Additional remarks:
I would like to remark on several observations concerning the **Space-Time Continuum** if it was real.

First, it would take a great deal of mass or equivalent energy to put a dent in the **Continuum** - if it actually existed. In other words, the **Continuum** is very rigid and not easily "stretched". For example, an energy amount equivalent to the mass of the earth would be needed to produce a one **"g"** acceleration, say on a starship. This level would be the long-term limit for onboard humans. An additional equal level of "negative" energy would be required to "zero out" the **"g"** field and "push" to achieve a greater effective acceleration—but in weightlessness. Therefore, a starship would need to generate energy levels equal to two Earth masses and the generator would have to operate continuously over long periods. This kind of generator is not a chemical or even an anti-matter engine that throws "stuff" out the back end; but would likely be a completely radical electromagnetic type of design. In any case, if the energy requirement is two Earth masses, and remembering $E=MC^2$, a tiny little starship's design and structure must be equivalent to two Earth masses and double this for the return trip.

Secondly, it takes huge amounts of energy to "stretch" just a tiny bit of **Continuum**, say a few light years, to reach the next closest star from our Sun. It is unimaginable what it would take to "stretch", "fold", "shrink" or form wormholes in the **Continuum,** as the case might be, to distant stars in our Milky Way galaxy let alone to other galaxies. The whole point being, "What is it really all for". Except for scientific interest, research and understanding our universe, there is no practical impact on our daily lives nor will there be for many, many, many years to come. Now, returning to the trampoline explanation of the **Space-Time Continuum**, it physically takes an energetic "push" to move any object to a different position. So, we pose this question:

"If it takes so much energy to move through the Continuum, why isn't there an energy exchange, i.e. slowing of the Earth in its orbit around the Sun??".

Chapter 2

"Run! The foundations are crumbling"

Chapter 1 pointed out that both **Special Theory Relativity (SR)** and **General Theory of Relativity (GR)** form important foundational inputs to the **Continuum** concept. That is why the implications of of the validity of these theories must also be examined. In **Chapter 1**, one serious flaw was exposed in the **Space-Time Continuum** theoretical concept. You may or may not agree with this finding; but Einstein himself stated the following regarding his **General Theory of Relativity (GR)**. He proposed three critical tests and if any one of these tests proved wrong, then **GR** would also collapse as a theory and so would the **Space-Time Continuum** concept.

1 Advance of the perihelion of Mercury.

2 Deflection of light by a gravitational field.

3 Gravitational Red Shift.

We will address these critical tests shortly in **Chapter 5** and show reasonable uncertainties and non-confidence in every one. But first, we don't really need to disprove any test or experiment, we need to only disprove Einstein's postulations or foundations that the **General theory of Relativity (GR)** and the **Special Theory of Relativity (SR)** are based on.

Postulation 1 (GR):
"There is no difference between a uniform gravitational field and an equivalent acceleration." Other wise simply known as the Equivalence Principle.

Postulation 2 (SR):
"Any ray of light moves with the determined velocity *c*, whether the ray is emitted by a "fixed" inertial emitter or by a moving emitter." *c* is a constant to all observers.

We will expound on these postulations and the existence of inherent flaws that invalidate the relativity theories, and consequently, **Space-Time Continuum** as well. But first, let's begin with an overall visual of the relevant foundations shown in **Illustration 9 "Space-Time Foundations"**. In this visual, Sir Isaac Newton's Newtonian physics is the foundation for force *(f)*, mass *(m)* and acceleration *(a)*. The formula *f=ma* governed celestial mechanics, gravity and trajectories from the late 1600's to the early 1900's when Albert Einstein entered the picture.

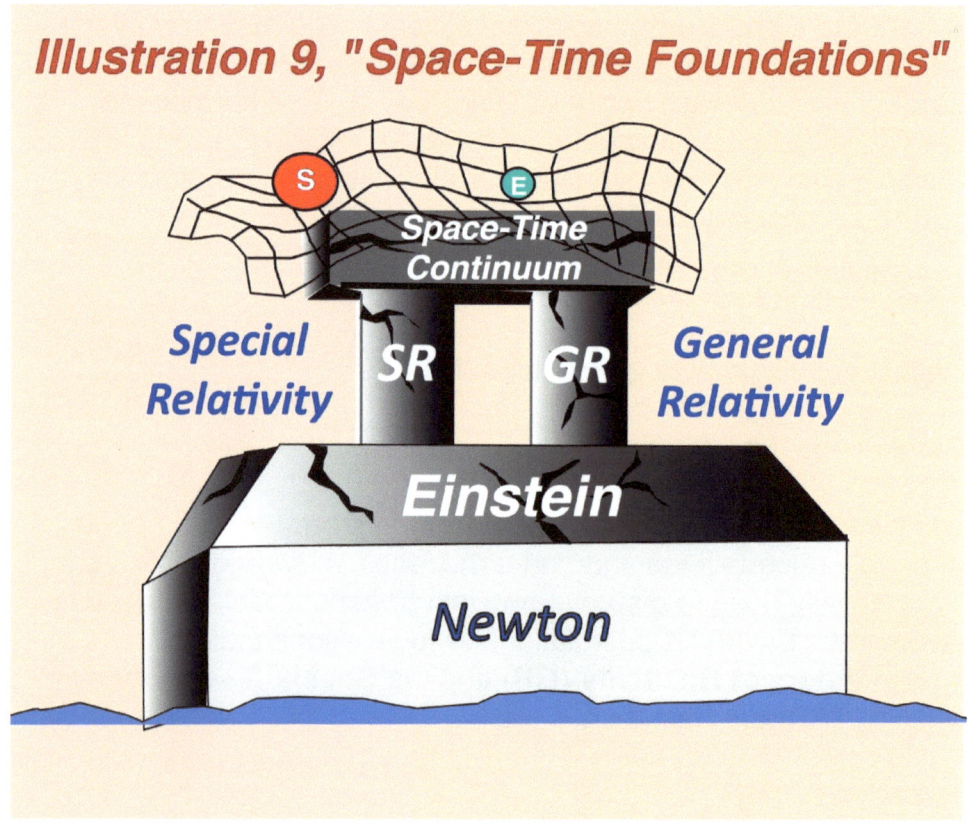

Newton's **gravitational theory (GT)** was considered as durable as granite and applicable for any velocity *(V)*. However, Einstein's **SR** extended **GT** to a supposed general form and made **GT** increasingly invalid for velocities *(V)* greater than about *0.2c.* To this day, there is controversy with the **Relativity** theories that "mainline" science has accepted to be absolutely confirmed and true. There are many dissident scientists and engineers, including this author, currently in opposition to the "mainline" community". But their voices and proofs of serious flaws go unheeded because "mainline" science is very comfortable with their firmly entrenched position. We will, later on, touch again on their reluctance to engage in serious evaluations and discussions from opposing Relativists.

Illustration 9 above depicted flaws in **SR** and **GR** that are presented herein, and that consequently lead to the crumbling and **shattering** of the **Space-Time Continuum** concept and the whole of **Relativity**. Before we identify the actual flaws, it is important to go back to the very beginning; i.e., the development of the governing postulations. pictorially shown In **Illustration 10, "Einstein Thought Experiments"**.

Illustration 10, "Einstein Thought Experiments"

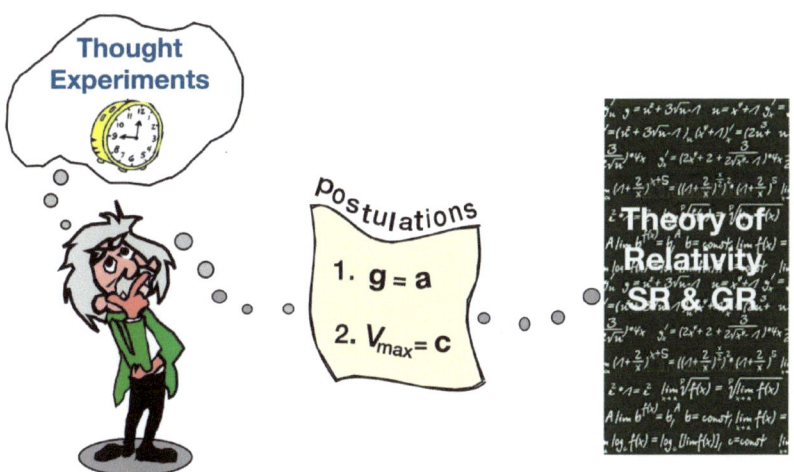

Einstein supposedly began with thought experiments before mathematically formulating a single equation. In these types of experiments, he visualized running along side of light beams, observing events from different observation locations, riding an accelerating elevator and a man falling from a rooftop. In his visualizations, he pondered the physics involved and deduced key governing principles leading to the statement of two famous postulations (assumptions). Sometimes they are erroneously stated in reverse chronological order as shown in **Illustration 10**. Postulation 1 was stated in 1915 while Postulation 2 was stated in 1905. Postulation 2, $V_{max} = c$, led to the development of **Special Relativity (SR)** and Postulation 1, $g = a$, led to the development of **General Relativity (GR)**. Basically, Postulation 2 says that the maximum velocity (V_{max}) that any object can attain is the speed of light (c). Postulation 1 says that acceleration (a) is equivalent to gravity (g); more commonly referred to as *"the equivalence principle"*.

As far as we know, Einstein did not undertake any preliminary analyses such as simulation of light ray paths or the kinematics of accelerating bodies, e.g., elevators and the like. Mathematical simulations of any sort would have been extremely time consuming and would have required maintaining rigorous precision. Computers were not yet invented which would have made these tasks possible in a short period of time.

It's my understanding that Einstein only had a short time, less than two months, from firming up his 2nd postulation **(SR)** to developing his mathematics and writing his paper for presentation. The title of his paper presented on June 30, 1905 was "On The Electro-dynamics of Moving Bodies". His **"eureka"** moment occurred when he stated his 2nd postulation which was in early May, 1905. It's hard to believe that he was able to mathematically develop **SR,** evaluate the meaning of his mathematics, author the paper, submit an abstract and prepare for presentation to recognized physics giants by mid-June, 1905. Somehow, I believe he was well underway with his manuscript before his eureka moment that "glued' his theory together. Why do I say this? Think about these unmentioned steps that also had to be taken during preparation of his manuscript. To avoid writing a fatal career ending paper and producing a "blooper" in front of his peers, he had to check, double check and maybe triple check each step of his mathematical development. At the same time, he had to compose and write his article, not only in a logical order, but with a quality literate style and with grammatically correct scientific prose. If you read a translated copy of his 1905 article, you would be impressed with its presentation quality. Although it is only my opinion, there's no question in my mind that it took time to "polish" and print his final article. All of these considerations certainly leave one to suspect that his mathematics and manuscript were nearly complete before he stated his Postulation 2. I leave the reader to his own opinion as we continue with the subject at hand.

Reference 1: A must reading
At this point I would like to reference a published book that is unique in identifying flaws in **SR** and **GR**. This short book does not rely on lengthy and boring mathematical proofs to uncover flaws, but rather by Illustrations and computer simulations. As they say, one picture is worth a thousand words. The book is easily readable and "digestible" for all ages ranging from early teens to the geriatric. *Reference 1* below is a major reference (in both paperback & eBook form) referred in this writing.

Reference 1:   *"RELATIVITY! ...Really? enter the Ping theory"*
                by Frank A Santora
                Published 2018
                Amazon.com book store

As mentioned previously, there are many current day dissidents and detractors addressing different areas of **SR** and **GR**. But, almost all apply derivation of complicated mathematics and physics way beyond the average "lay" person. The problem with this approach is that their work can never be conveyed to the average citizen who is not skillful in these science areas. Another problem is that it is very difficult and time consuming to examine another person's work for the purpose of confirming its correctness. Most scientists including both dissidents and the mainstream scientists cannot afford to dedicate time to anything other than their own work and pursuits, especially mainstreamers. Thus. the two camps remain unreconciled and separated. **Reference 1** bridges the two camps and also bridges both camps to the general public. Essentially, **Reference 1** can serve as a pre-requisite or a post-requisite to this book **"Space-Time Continuum ...Shattered"**. With this preamble we can now proceed to **Chapter 3, *"ALERT! Special Relativity Flaws Ahead!"***

Chapter 3

"ALERT! Special Relativity flaws ahead"

Reference 1 introduced a unique "Ping" analysis concept that uses one (1) wavelength of light with its particular quanta (photons) to trace the true path that observers actually see. This engineering computer analysis approach included looking at light paths and associated Doppler effects on stationary and moving observers and their interpretations of such effects. Doppler effects and specific light quanta have not, heretofore, been considered in trying to understand proposed "thought experiments" that led to **SR** and **GR** theories and related paradoxes. I strongly recommend that you obtain and read **Reference 1** as a Pre- or post-requisite to this book. It is available both in e-book and paper back form.

Reference 1 examined the *"The moving train paradox"* and the *"Simultaneous lightning strikes paradox"* that help form the basis of **SR**. **Reference 1** revealed that by selecting the correct light path and its specific quanta and determining what the stationary and moving observers actually see, startling outcomes result.

ALERT! <u>Flaw #2</u>, *"Appearances are deceiving"*
The results proved that what the human mind 'thinks' it perceives as a straight path is really an ***apparent curved path***. Now, when this deceptive curved path is used in Postulation 2, the mathematics development that must retain certain science truths (i.e., speed of light **(c)** is constant in a vacuum) leads to a strange world. In this new mathematical world (see **Illustration 1**) lengths, clocks, and masses must change in order to "straighten out" the curved light path. Furthermore, a mathematical "singularity" point appears at the speed of light **c**. Everything in the mathematical world "explodes" at this point. Hence the speed limit **c** applies to any object attempting to attain this speed. However, the "Ping" analysis approach ***returns the physical world back to normal*** as we know it —no physical speed limit!,... (whoo! ..what a relief!)! In the continuing discussion below, we will further confirm this and

examine another Einstein thought experiment, namely; *"Chasing a light beam"*, and exposing another flaw in Postulation 2.

Chasing a light beam

Another of Einstein's thought experiments imagined himself running along side a light beam as shown in **Illustration 11, "The Case Of The Frozen Wave "**. Looking first at the right side of this illustration, we see Einstein running slower **(V < c)** than the speed of light **(c)**.

I want to pause here for a moment. In all the illustrations presented herein, I mean no disrespect toward Dr. Einstein. Depictions of him are meant to only add a touch of light-heartedness to a serious subject.

Illustration 11, "The Case Of The Frozen Wave"

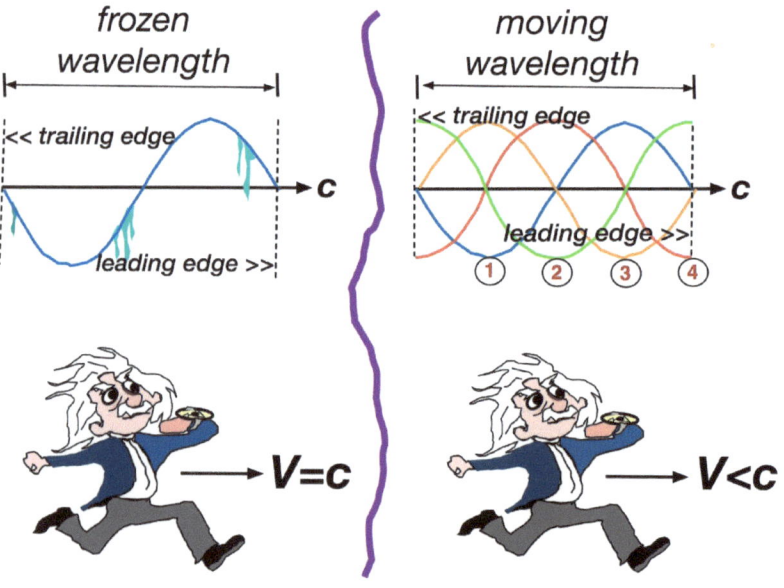

I believe that his contributions of the **Theory of Relativity**, flawed or otherwise, have been the impetus for scientists to think "outside the box". Many new areas and theories in Astrophysics have already been proposed and many are currently being researched. It is important to note that Dr. Einstein has contributed valuable contributions to science in areas other than **Relativity**. He himself injected humor many times regarding his work, his personality, and the work of his colleagues. I truly believe that his work will lead to other exploratory avenues that would never have been foreseen, and these in turn will result in significant, momentous discoveries and advancements.

Now, let's turn our attention back to chasing a light beam. When Einstein pictured himself running slower than *c*, he saw a beam of light moving past him and progressing through positions 1, 2, 3, 4 before repeating itself. He looked at his wrist watch and compared it with the beam source that was emanating from a distant clock behind him. It appeared that the clock behind him was slowing down. He ran faster until he was running at the speed of light *(V=c;* left side **Illustration 10)**. Now as he looked at the light beam, it appeared to be frozen and not oscillating. Further, the distant clock image in the light beam had come to a complete stop. He reasoned that light waves cannot freeze! Other physics laws appeared to be violated as well. I believe this is what Einstein was "held up" on in his **SR** theoretical development and was unable to make sense of it until his "**eureka**" moment. Hence, he went on to reason that light waves arrive at any stationary or moving observer at the speed of light *(c)*. Thus Postulation 2 was born.

We begin finding the **Flaw #3** in Postulation 2 using **Illustration 12, "Relativity Doppler Error"**. In this illustration λo represents the single wavelength of a continuous wave that is being transmitted at the speed of light *(c)*. Let's assume that the transmitted wave is "yellow" with a wavelength $\lambda o = 570$ *nm* (nanometers), which is equivalently equal to 0.0000225 inches long. You can see that, if an observer is stationary *(V = 0)*, the observer will see the same color as transmitted, namely; yellow (570 nm). However, if the observer is moving with some velocity*(+V)* away from the light source, the light ray's effective impingement speed wavelength *(λ')* (detected or observed} is not *(c)*, but *ceff (c minus V)*. This lower than *c* effective speed requires a longer time increment *($\Delta t''$)* to detect the full quanta bundle of photons. This longer time is the result of Doppler frequency shift *($\Delta \lambda$)* that results in a longer or "stretched" detection. The doppler shift is calculated by the straight forward equation $\Delta \lambda = (V/c)\lambda_0$ in **Illustration 12**. Note that this equation is completely well behaved even for speeds **(V)** greater than **c**. I would like to comment here that the Doppler effect and its compensation is actually experienced every day with acoustic, radio and video wave transmissions and detections. It is a very real common phenomenon in our daily lives.

Illustration 12, "Relativity Doppler Error"

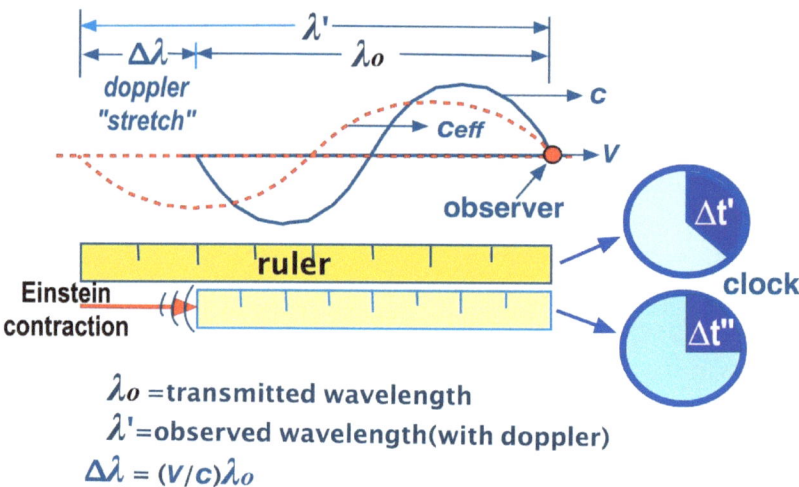

λ_o = transmitted wavelength
λ' = observed wavelength (with doppler)
$\Delta\lambda = (V/c)\lambda_o$

In **Illustration 12.**, two rulers or measuring rods are shown. The upper ruler measures the length of the true observed wavelength **(λ')** while the lower ruler measures the length of the original transmitted wavelength **(λ_o)**.

ALERT! Flaw #3, "Contraction contradicted"

Now, Einstein states in Postulation 2 that an observer cannot detect whether he is moving or stationary. What he sees in a stationary frame is what he **MUST** see in a moving frame. Therefore, in **Illustration 11** at a speed of **V=c**, the observer is **forced** to see the original transmitted color, say yellow (570 nm) and not the observed color red (1140 nm). This is **Flaw #3**! He does not account for Doppler shift. Not accounting for the Doppler shift mathematically forces the length of the bottom ruler to shrink or contract by the amount equivalent to the Doppler shift **$\Delta\lambda$**. This is a famous result of **Special Relativity** and is referred to as Einstein's **length contraction.** Now the length of this fictitious ruler compared to the length of the true upper ruler is λ_0/λ' which is equivalent to any measuring rod or object length **(L')**

$$\lambda_0/\lambda' = L_0/L' = 1/[1+V/c] \text{ or}$$
$$L_0 = L'/(1+V/c)$$

Again, notice that there is no singularity. If the velocity of a moving observer, for example, is **0.25c**, then the length of $\lambda_0=L_0$ is 80% of the the observed or measured length. Therefore, if an observer is traveling in a Starship measuring 1000 feet in length and moving with a constant uniform velocity **V=0.25c,** his apparent Starship length is 800 feet because of **_Flaw #3_**. **SR** theory and its attendant mathematics is essentially dictating what he must see without accounting for Doppler effects. Let's carry this thought further and assume the observer travels at **V=c** and four times the speed of light **V=4.0c**. At the speed of light **(V=c)** his apparent Starship length is 500 feet and at a "warp speed" of **4c**, its length is 200 feet.

The same equation expressed by Einstein is given below.

$$L_0 = L'[1-(V/c)^2]^{1/2}$$

Einstein's **SR** mathematics implies that the length contraction at **V=c** is 100%; i.e., the Starship length shrinks to zero feet and consequently cannot go faster than the speed of light. Right about now your eyebrows are raising. We have two length contraction equations, and both implying that a physical length shortens with increasing speed. These equations are certainly metaphysical in nature and are the cause of great consternation. The length contraction occurs only in the direction of velocity **V**. Width and height dimensions of an object do not change since they are orthogonal to the direction of **V**.

Other metaphysical questions arise. For example, "Is the object's volume "squashed" or "densified" with the shortening?". "Does the object's atomic and molecular lattice disintegrate?". How about heat buildup?", etc. What happens when the observer is approaching the source **(-V)**? Does a physical mathematical singularity present itself?

There had to be some other factor or factors involved that would make this outcome plausible. In my opinion, Einstein was already well along in his mathematical derivation when this "snag" was encountered. He faced a very difficult conundrum. On one hand, his **SR** theory predicted a physical shortening of length with a moving observer. On the other hand, his Postulation 2 states that a moving observer cannot detect motion and that all objects appear as if at rest, **V=0**. Something had to give! Enter time as the *fourth* dimension in the **Space-Time Continuum** concept.

Enter time dilation
Returning to **Illustration 12**, we see two clocks. The upper clock is an onboard clock that travels with the onboard observer moving at speed **V**. This clock measures the actual or true time **(Δt')** to detect a full single wavelength of light as affected by Doppler shift. The lower clock indicates the time **(Δt")** for a moving observer, at speed **V,** to detect a wavelength that has no Doppler shift. How does Einstein reconcile these two different clock readings for the same event? He's facing the same dilemma as he did before. Both readings cannot be right, but he reconciles them by introducing the concept of "**time dilation**" into the theory. Time dilation simply means that it takes longer to observe an event when the observer is moving with velocity **V**. This, I believe, is another **eureka** moment that "glued" together his theory of **Special Relativity**. Einstein's time dilation equation is as follows and is just the inverse of **L₀/L'** above.

$$\Delta t''/\Delta t' = T''/T' = 1/[1-(V/c)^2]^{1/2}$$

If we take the same example as above of **V=0.25c** we find that the time dilation is a factor of 1.033 of the rest time. The corresponding factor for length contraction is 0.967. Thus, the relativistic time increases when the relativistic length contracts. In effect, this offsetting compensation, knowingly or not, corrects the omitted Doppler in Postulation 2; but still, the length contraction and time dilation must be accepted as real physical phenomena. The author simply views this correction of time dilation and length contraction as mathematical artifices that are happenstance toward the "correct" solution. Actually, the two Einsteinian equations are the same identical Lorentz equations as referred in **Reference 1**. I don't know whether they are the natural outcome of Einstein's derivations or a useful set from Lorentz that was utilized in the derivation of **SR**.

ALERT! *Flaw #4*, *"SR singularity singularly wrong"*
Reference 1 proved that the Lorentz transformation equations are not required in solving for actual path lengths of light rays or their travel times; thereby *invalidating any need for applying SR*. Further, it was shown above what is believed to be an erroneous **SR** development that leads to singularities, unreasonable artifices and serious limitations in the theory. Therein, length contraction and time dilation are debunked as only apparent and abstract illusions. I would like to mention, as an after thought, another abstract outcome of **SR,** and that is the increase of mass as an object's speed increases. Again, **SR** predicts that mass increases to infinity (∞) as *V* approaches *c.* Can you imagine what happens if a planet or a relatively small Starship approaches the speed of light? It would become so massive that it would be basically a so-called *black hole* and would gobble surrounding galaxies. Unimaginable! And why? ...because of the singularity present in the **SR** theory.

This brings **Chapter 3** to an end, although much more can be discussed, this is sufficient for the scope of this book. In the following **Chapter 4**, we will address the flaws of the second foundational pillar that supports the **Space-Time Continuum** concept.

Chapter 4

"ALERT! General Relativity flaws ahead"

This Chapter looks at the *"Moving Elevator Paradox"*, or equivalently, the *'Equivalence Principle'* that forms the basis of **General Relativity (GR)** from Einstein's Postulation 1.

Einstein recognized that his theory of **Special Relativity (SR)** was quite limited. It did not include acceleration or gravitational effects in his quest for a more general inclusive theory. Einstein, when considering this dilemma and knowing that if his **SR**, published fifteen years earlier, was to remain valid had to be extended to include acceleration *(a)* effects. I again get the feeling that Postulation 1 was now the "glue" needed to make the theory development plausible. In other words, if he could prove that acceleration equals gravity which in turn is an attribute of mass, then he would have a plausible theory. This more comprehensive theory is now referred as the **General Theory of Relativity (GR)**. The *moving elevator paradox* supposedly did prove that $a = g$. "How?" you might ask. I might say, "By the Equivalence Principle!". Einstein, based on this principle, could now substitute *g* for *a* and declare that a large mass, such as the earth *(1g)*, is the reason that a curved light path is the result in the accelerating elevator. However, in the end, this is an incorrect statement!

This chapter demonstrates the difference in accelerations between an elevator at-rest on the ground **(1g)** and an elevator that is accelerating 'upward' at **1g**. The reader is strongly encouraged to refer to **Reference 1** given in <u>Chapter 2</u> above, for the full explanation of the flaw found therein. Proving that trajectories inside the elevators are NOT the same is a startling result. The difference, although subtle, raises a serious challenge to the validity of using the *Equivalence Principle* as an underpinning to **GR**. Forcing the presence of a gravity field and light beams to bend in an acceleration environment certainly would lead to mathematical formulations that produce non-intuitive and imaginary worlds.

The Equivalence Principle"

Illustration 13, "Moving Elevator Paradox" places a man in an enclosed room or elevator that is at rest on the Earth's surface in a 1**g** gravity field. He drops a ball that <u>accelerates</u> to the floor due to gravity. If it is an elastic impact, the ball will rebound to its original drop height and its trajectory is repeated continually. Of course, this assumes that there are no other dissipative forces, such as air drag, present. **Illustration 13** depicts this scenario with the elevator on the left. Now the paradox also places a man in an elevator with

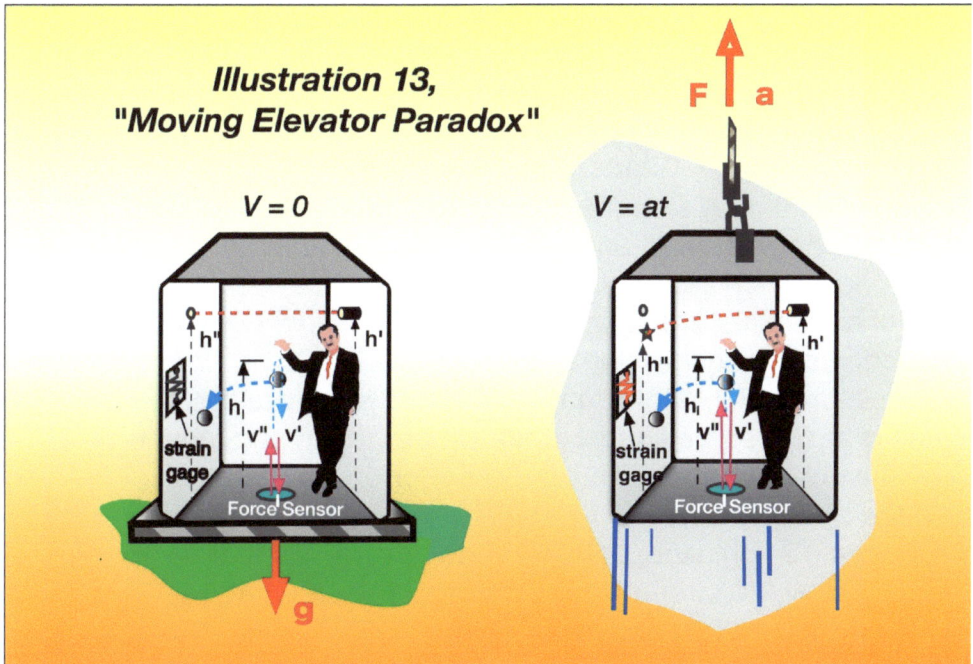

acceleration **a**. Einstein envisioned a sky hook attached to the top of the elevator pulling with a constant force and producing a constant acceleration equal to gravity (**a=g**). This is depicted on the right in **Illustration 13**. This *"moving elevator paradox"* poses this 'thought experiment', that further stated that the man in the elevator could not distinguish nor determine that he was in an accelerating elevator. Why, because when he releases a ball it follows the identical rebounding trajectory as in a 1**g** environment. As far as the man was concerned, he could well be at rest on Earth's surface and dropping the ball

Einstein then presumes, according to postulation 1(see Chapter 2), that the physics occurring within the accelerating elevator must be the same as the at-rest elevator. Einstein and many others are right, up to a point, that the relative trajectory of the ball is exactly the same in both situations. However, their conclusion that both observers in the elevators are occupying the same physical environment is faulty. In the at rest elevator, the ball is experiencing an accelerated downward trajectory. However, in the accelerated elevator, the ball is not being accelerated and therefore is in a constant velocity "free-fall" trajectory at the moment of release. There are two red flags that signal concern and should have been examined in more detail:

ALERT! *Flaw #5, "Deceiving bouncing balls"*
The accelerations (***g*** & ***a***) are different in nature and direction. Gravity ***g*** is an "***internal***" acceleration field. In other words, the field acts individually on every molecule and atom inside an object. The matrix of molecules and atoms that form the object do not depend on 'mechanical' transmission of forces to react. So, in **Illustration 13**, left picture, every individual atom of the structure and the man senses ***g** **at the same time***. He feels it as weight and the elevator "feels" it as compression. This is not so for an acceleration produced by an "***external***" force **F**. In this case, referring to **Illustration 13** right side, the force **F** is transmitted to the elevator structure. Only the strength and rigidity of the structural components, e.g., rivets hold it together. The man standing inside feels a reactive force or pressure transmitted through his shoes to his feet. Thus, his whole being is accelerated via connective bones, ligaments, and tissue. When the two men stand on a force sensor, say a weight scale in their respective elevators, they will weigh the same. Apparently, man's senses are not keen enough to sense, in this case, the difference in forces acting on or within him. But, if the force sensor measures the impact force of the dropped balls, the impact force in the accelerating elevator would be twice that of the at-rest elevator. That's why it's important to look a little deeper rather than relying on human perception and logic alone to form a formal statement of fact. Simply stated, we have shown in the above discussion, trajectories of dropped balls inside the two

elevators are not the same but they are *deceptively* the same! The proof lies in the subtle difference in the following statement.

"In the at-rest elevator, the ball is hitting the floor; but in the accelerating elevator, the floor is hitting the ball."

Therefore, *the Equivalence Principle is NOT valid!* Now we can move on to another concern, light propagation.

Additional Concern
GR states that the path of a light beam is bent by the curvature of the **Space-Time Continuum** due to the presence of a massive celestial body (e.g. the Sun). Einstein extended his moving elevator thought experiment to use the *Equivalence Principle* to predict the effects of a massive body on light propagation.

Referring to **Illustration 13**, we see the men in the elevators lobbing the ball toward the opposite wall. Again, Einstein visualized that the trajectories of both balls curved downward thus confirming his Equivalence Principle. As pointed out in the discussion above, the downward observed trajectories appeared to be equivalent. Einstein then extended his visualization of a tossed ball to a photon in a light beam. He reasoned that a light beam would be deflected to some degree when propagating through a strong gravity field. Not only did he not recognized **Flaw #5**, he did not acknowledge that the ball is a mechanical or kinematics event and a light beam is an electromagnetic event. The momentum in the accelerating elevator is imparted to the ball, but is not imparted to an electromagnetic source event. In effect, he was mixing "apples" and "oranges" in his theory.

ALERT! Flaw #6, "False-Negatives=Proof-Positive"
Referring again to **Illustration 13**, a laser light source is mounted on the wall of both elevators. If light pulses are 'shot' to the opposite wall in the at-rest elevator, they will impact the wall at the same height **h** as the red laser source. Is there any gravity 'bending' of the pulses? I believe the answer is yes. Physics tells us that photons, although massless, do have a mass-energy related

attribute and therefore will exhibit some bending. But any bending is due to the massy attribute of photons in a gravity field and NOT to **GR**! This bending has been predicted with other theoretical approaches which are beyond the scope of this book. How do we prove this statement here?

We prove this statement by making two "false-negatives" assumptions to arrive at a "proof-positive" outcome. Our first false-negative assumption is that there absolutely is no bending of a light beam from any phenomena in the at rest elevator. This assumption statement is depicted in **Illustration 13** in the at rest elevator. It shows a laser source mounted on the elevator wall at a height **(h')**. The laser path (shown in red) is a straight horizontal line that impinges the opposite wall at a height of **h''**, which is the same as the source height. Next, the same false-negative assumption is applied to the accelerating elevator. But, in this case, the man sees a downward laser path that strikes the opposite wall at a height lower than the source height **(h')**. This makes no sense since the laser path is not, by dictation of our false-negative, influenced by gravity. So, why do we see bending of the light path? The curved path is due to the upward movement of the elevator during the travel time of the laser path from one wall to the opposite wall. The answer is simply factually attributed to the relative motion of a moving elevator. The curved path has nothing to due with the equivalence principle or **GR**. And there, my friend, is the "proof-positive" that the Equivalence Principle is incorrect and flawed! The result is that the foundation of **GR** is seriously, if not fatally, flawed!

Chapter 5

"Oh, the error of our ways"

The previous Chapters pointed out a number of flaws in the theories of **Special Relativity (SR)**, **General Relativity (GR)** and the **Space-Time Continuum** concept. It is only natural that the reader might ask the following question:

"Haven't there been a number of experiments conducted that proved Einstein right?"

Before we answer the question, we have to place the whole of this subject in proper context. The context involves personal philosophies, stances, idolization, and pursuits of distinction; not unlike political parties. As mentioned in **Chapter 2**, controversy exists today regarding **SR** and **GR**. Currently, the mainline scientific community (Group A) supports and energetically advocates the acceptance of Einstein's theories and concepts as factual. However, there is another lesser known community (Group B) and they do not accept **SR** and **GR**. *Illustration 14* pictures this current stance of the two communities or groups. Why is there such reluctance of these two groups to share a dialog and/or to debate the involved issues? Well, there are a number of reasons why and they are becoming less resolvable with each passing year, unless an experiment(s) or test(s) can be conducted that irrevocably proves either Einstein correct or wrong.

Opposing Relativists
First, let me give you where I'm coming from. This author has been involved with the **Relativity** theories for only two short years as of this writing, and only as a spectator. I am not a member of any organization or group. However, I did author a brief book in 2018 — namely; *Reference 1* given in **Chapter 2**. *Reference 1* delineated my concerns about the validity of the initial postulations which formed the foundations of the theories themselves. During this time, I've observed and experienced certain notable happenings that have impressed me and have led me to form an opinion of the two

aforementioned groups and their world of **Space-Time Continuum**. It is this opinion that I am expressing here. Being an octogenarian and well retired, I have no purpose for monetary gain or recognition except to be acknowledged for opening this investigative avenue. Perhaps this book may be a starting point for discussion and debate between the two groups.

Illustration 14, "The Twain in Vain"

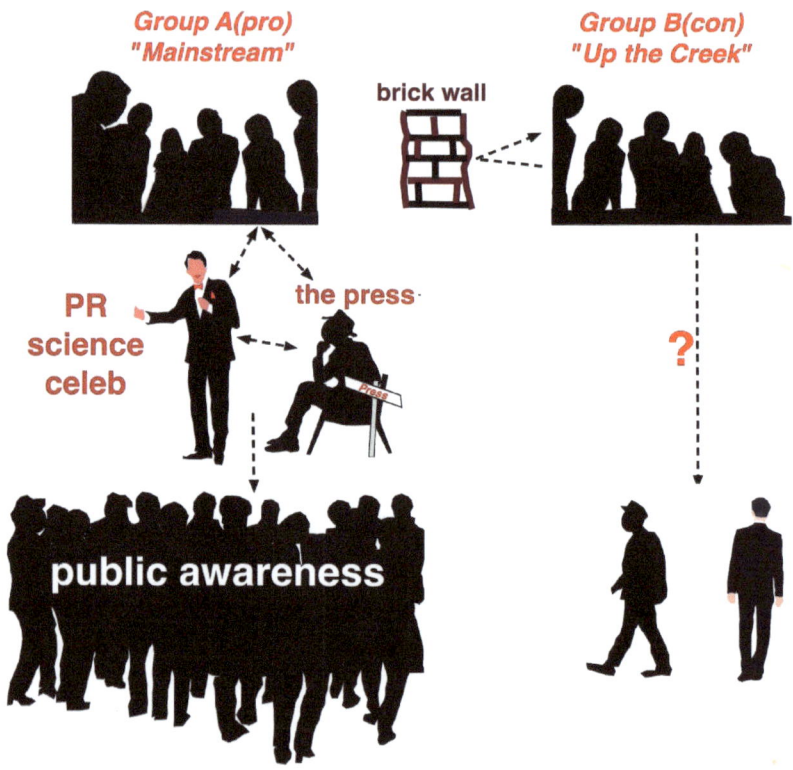

The top left side of **Illustration 14** shows Group A scientists who advocate and support Einstein's theories of **Relativity**. They have rederived and checked Einstein's physics and mathematics. This group is made up of physicists and engineers who have designed and performed experiments to validate the theories. These experiments go back many years from the times of Einstein up to the current time. They range from normal laboratory type experiments to very sophisticated satellite programs. Today, universities teach **Special Relativity (SR)** and **General Relativity**

(GR) courses as, for the most part, validated theories. As far as I'm aware, little or no opposing and contrary positions are presented for student instruction. This educational influence seems to continue into their research and livelihood life. Thus, it is only natural that mainstream science is reluctant to consider any anti-**Relativity** sentiment that would be looked upon unfavorably by their peers and employers. This stance results in two outcomes. First, a sort of wall is "built" between the mainstream science community and the Group B community— what I refer to as the "up-the-creek" community. Secondly, I detect an undesirable animosity and arrogance between the two global groups. The reason for these feelings is that one side is convinced that **SR** and **GR** have been been proven valid. Therefore, in their sight, Group B consists of "wayward" scientists and engineers and any contact should be avoided.

There are other aspects of this author's opinion. Group A generates many papers, articles, documents, videos and reports of their research and experiment findings. They are very mindful that any reported finding that validates or extends the **Space-Time Continuum** concept will, almost instantaneously, lead to peer recognition and maybe fame as well. So, there is great competitiveness to be "first". This is not bad. It's certainly good for one's ego, self esteem and financial gain. But in this environment, it is extremely difficult to admit failure on some research item or project.

Now, both groups have a serious problem. Most all of their endeavors and reports are "encased" in complicated layers of physics, mathematics and engineering analyses. Only peers in the groups can make sense and importance of their endeavors. Thus, no one in the general public, without these science skills, can understand or fathom what it's all about. The solution to this problem is what I refer to as the "middle" man. ***Illustration 14*** portrays these men or women as a "Public Relations (PR) science celebrities". These people are very special people with extra special skills and talents. They are adept in speaking and articulating their thoughts in understandable language to any layman. They generally have a Ph.D. in a science or physics field and also have a working understanding of the involved sciences. They are very genial, acceptable in appearance, and likeable to any age level. In other words, he or she has the "total package" to be the perfect conduit between scientific groups and the general public. I'm sure you have seen these PR people deliver and expound on the **Relativity** theories in an entertaining and informative manner. They have appeared in many TV documentaries, interviews, magazine and newspaper articles. They have also published many books that are easily available to the general population. Yes, thanks to these PR celebs, the general public has an awareness and some understanding of **Space-Time Continuum** and the world of Einstein's **Relativity**.

We also see that the media and press frequently seek out these PR middle people whenever there is news in the Group A community, but the press and media seem unaware of the existence of Group B as depicted in **Illustration 14.** Consequently, I have called Group B the "up-the-creek" community since they have no similar PR science celebrity or press presence to relate to the public. Group B is almost a closed entity with communication chiefly amongst themselves. Any public awareness of their endeavors is by happenstance, namely by "passersby".

With the above discussion, we can now continue with the subject of this Chapter, namely to answer the question: "Haven't there been a number of experiments conducted that proved Einstein right?" But before proceeding, the following discussion may be too boring or too complicated for most readers. Therefore, it's OK to skip to **Chapter 6**; the general gist and topic of this book is not diminished.

Has Einstein been proven right? My short answer is "No!" Although Group A has presented results of tests and experiments as proof, there is great consternation and unacceptance of their findings within Group B. Remember, Group B is also composed of high quality and highly skilled scientists and engineers. Their expertise and findings should not be prejudged "out-of-hand" as "wayward" and "crackpot". **Illustration 14** shows that communication attempts of Group B with Group A are completely ignored. Group A, it seems, has an unconscious desire to protect the theories of **Relativity** at all costs thereby elevating the theories idol status. Maintaining this idol in the public eye and staying entrenched therein, Group A enjoys a very envious position favorable to continued recognition, fame, celebrity and financial gain. There is nothing to lose in their lifetimes or generations after them until an experiment or round-trip manned mission can be performed at relatively high speed, say about **0.2c** to absolutely prove whether the twin paradox is true or not. This author believes that, until that future time, any experiment or test will be clouded with uncertainty, because of design error sources, extremely small "Einsteinean" verification parameters, unknown error sources, alternate application of physics principles, etc.

Elaborating further, the proofs of such experiments to date, depended on determining very small "Einsteinian" parameters connected with his theories. Unfortunately, every experiment performed thus far, has many inherent error sources as small as the parameter that is being determined. Error sources (static or dynamic) can come in the guise of systemic, measurement, design, observation and even unknown forms. And therein lies a major, major problem—namely, any unaccounted error source or inability to derive the actual magnitude of the error source can be interpreted as the theoretical parameter being sought. In other words, the parameter value sought can be buried in a sea of potential errors, and *if the theory is actually invalid, any unknown or insufficiently defined error source can erroneously validate the theory.* It is noted that error sources can be debatable and maybe even invented. Also, these experiments are very expensive and can involve many highly skilled personnel and a lot of time. It is also very important to recognize that it's literally impossible to revaluate any experiment that has already been performed whether in the past or near present. "Why?" you might ask. Well, all the original equipment or assembly is no longer intact or operational. Neither are the original measurement data or their related processing available. Replication of the test and experiment equipment would never be of the same manufacture and assembly quality as the original and, therefore, out of the question. All the original error sources of the experiment would be different in magnitude, measurement, and many additional unknown sources could easily be introduced. Any current day measurements and processing with updated equipment would definitely result in different calibrations and data reductions. It is better to start from scratch to repeat an old experiment or to perform a new one. These statements will be impressively evident when we consider the multi-million-dollar Gravity Probe-B experiment below. Most experiments or tests have been challenged by Group B, and no debate or resolutions of Group B challenges have been forth coming from Group A. The various experiments remain in limbo. Therefore, in the following discussions, the unanswered concerns of Group B are noted by a "yellow" caution flag:

We now turn our attention to the three experiments associated with the success or failure of **Space-Time Continuum** as expressed by Einstein himself (see **Chapter 2**), namely;

1. Advance of the perihelion of Mercury
2. Deflection of light by a gravitational field
3. Gravitational red shift
4. Gravity Probe-B

A fourth experiment, Gravity Probe-B (GP-B), is included as well. This latter is, perhaps, the most costly, technically challenging, and time consuming experiment of all.

1. Advance of the perihelion of Mercury
In this experiment, an anomaly was observed in the advance of Mercury's perihelion around the Sun. ***Illustration 15*** shows the definition of perihelion advance with exaggerated elliptical orbits. Initial astronomical observations of Mercury's orbit began in the 1700's and continued into 1800's. About 1859 an anomaly in predicting the precessional perihelion advance of Mercury was noticed. Perihelion is the closest approach of Mercury to the Sun. However, the location of perihelion is not "fixed". It is continually moving in a counter clockwise direction *(ϕ)* relative to the Sun as shown in ***Illustration 15***. What was noticed was that there was a disagreement between what was mathematically predicted and what was observed with actual telescope measurements. This difference amounted to about 38 arcseconds (asec) per century or about 0.01 degrees per century. To put this in better perspective, the difference of 0.0001 degrees per year was unaccounted for.

You may ask, "Who cares about such a "piddling" amount?" Well the scientific community cared, because it set off a quest to find the cause of this discrepancy. Hence, enter Einstein and his **General theory of Relativity (GR)**. The answer appeared to be an omission of the **GR** contribution of 42.98 asec per century to the Newtonian predicted value. Over passing years, observational data as been updated to a currently accepted value of 574.10 asec per century. Also, the non-relativistic Newtonian value was updated with refined models to a currently accepted value of 531.63 asec. per century. So, now the difference between observed and predicted is 42.47 asec per century. The predicted **GR** contribution is 42.98 asec per century, thereby now leaving a discrepancy of only 0.51 asec per century. You might now say, "OK, problem finally solved!"; Let's move on to bigger and more interesting things". Before we move on, let's consider a few more interesting facts. One fact is the uncertainty in the calculations for 574.10 asec and 531.63 asec. The result of considering these inaccuracies indicate that the range of minimum and maximum values of the observe difference in perihelion advance is 41.10 asec to 43.81 asec respectively. In turn, these inaccuracies still result in a range of acceptable values.

Illustration 15, "Precessional Advance of Mercury"

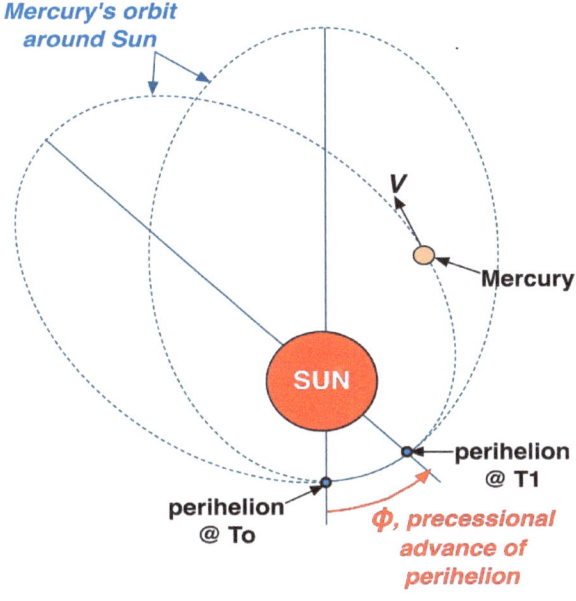

that support **GR**'s validity. But wait a minute, let's consider the second fact that we are in the computer age.

CAUTION! 🚩 Computer age ahead

Before the computer age of the 1960's, all celestial orbit analyses had to be performed by laborious hand calculations. Further, these manual calculations assumed certain simplifications of the solar system to ease the burden. For example, all planetary orbits were assumed to be circular except for Mercury, and all planet orbits, including Mercury, lay in the same plane. With the advent of computer technology, these laborious calculations could now be easily performed and with a high degree of precision. Now it became feasible to eliminate any simplifying assumptions and omissions. In fact, research was accomplished in 2003 and published in "Alternative Physics" by Bernard Burchell. The referenced paper can be found at the following web site:

http://alternativephysics.org/index.htm

The paper reports the results of a computer numerical analysis that uses Newtonian gravitational masses of the planets and Sun in determining the Newtonian accelerations, resulting velocities, distance and perihelion displacements. The ephemerides (positions in time) of the planets do not include any relativistic adjustments. By taking a numerical integrating approach, the actual planet eccentric orbits and their rotating inclined planes yield very accurate Newtonian motion over centuries of time. Thus, a full 3-dimensional computer simulation of the simultaneous movement of the entire solar system is possible including movement of the sun about the solar system's barycenter. The simulation results showed that the Newtonian value of perihelion advance is 528.25 asec/century. This value is less than the observed value of 574.10 asec/century by 45.85 asec./century. Depending on the accuracy of the observed advance, adding the **GR** contribution of **43** asec/century still falls short by 2.2 to 3.5 asec/century. This sounds trivial but if the Newtonian prediction is correct, then the **GR** contribution is short by an unacceptable 7%, thereby indicating that **GR** is unable to account for the shortfall. There are other concerns with the simulation completeness. Were other masses included; such as the Earth's moon, asteroid belt, etc.? What is the contribution of solar radiation pressure on Mercury? It seems to this author that a fresh new experiment repeat is needed. However, this time, two independent research teams should be used; one from Group A and another from Group B. It would be more expensive, but it just may have a chance for resolution.

2. Deflection of light by a gravitational field

The second experiment we want to briefly look at is the more well-known light deflection by a massive body. This experiment involved observing starlight rays passing the Sun during a lunar eclipse to detect ray "bending". If bending or deflection can be observed, then Einstein's **General Theory of Relativity (GT)** is proven valid. *Illustration 16* depicts this experiment endeavor showing an exaggerated light bending effect.

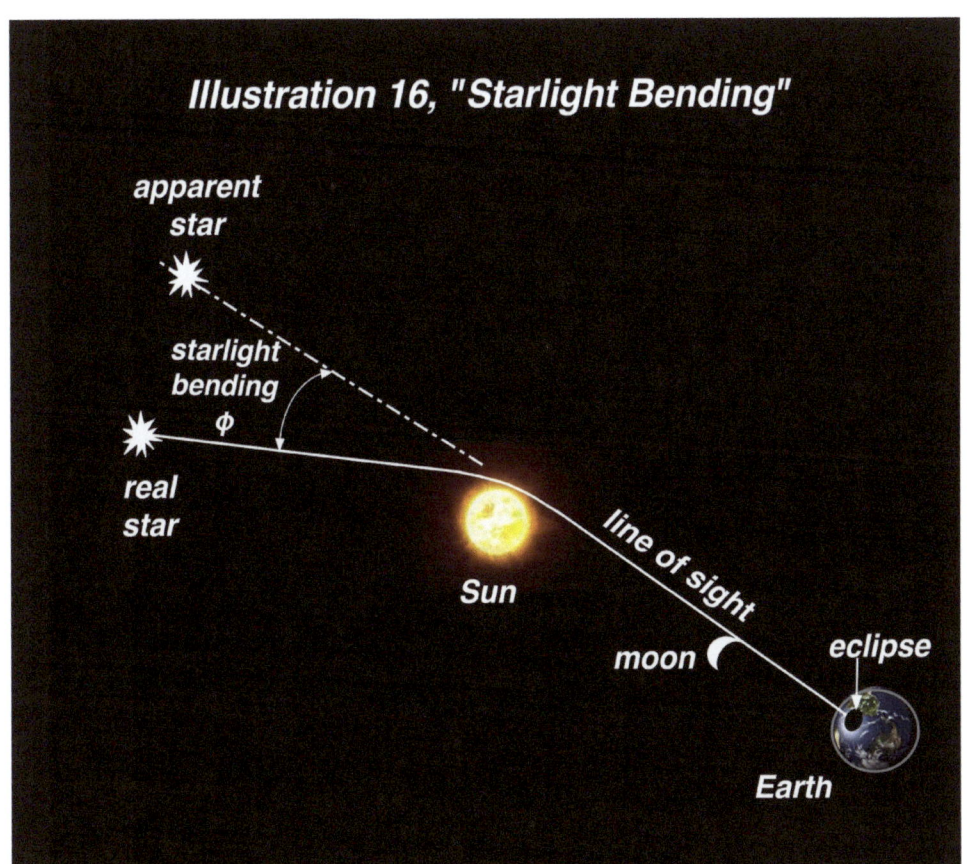

In this endeavor, the experiment goal was to detect and verify Einstein's **GR** prediction of the bending of a light beam by a massive body. In this experiment, the massive body is the Sun and the light beam is a distant star. The exaggerated bending of a light beam is denoted by ϕ in the illustration. Actually, the experiment can only be accomplished during a total eclipse of the Sun by Earth's moon; when it is possible to observe stars close to the Sun's limb.

In order to get the real feeling of the difficulty to accomplish this detection of starlight, we must look at ϕ again. The magnitude of this "Einsteinian" parameter is a mere 1.75 arcseconds (arcsec) or equivalently 0.000486 degrees. Wow! Therein lies the challenge; to measure this parameter in the presence of known and unknown error sources both larger and smaller than this predicted value of 1.75 arcsec. The quest to take on the challenge began back in the early 1900's and has continued to this day, the latest being in 2017. The reported 2017 experiment has received little publication. It has been reported that it agrees with the 1.75 arcsec ±.0.06 arcsec.

The Table below shows the reported results of conducted experiments rom the 1900's to 2017. At first glance, the observation results show close agreement with the Einstein prediction of 1.75 arcsec. However, there are serious non-convincing aspects of these findings. Although the mainstream scientific community (Group A) has accepted these results as proof of Einstein's **GR**, Group B has major misgivings.

CAUTION! 🚩 *Misgivings around the corner*

TABLE 1 below provides the results for a series of experiments ranging from 1919 to 2017. First off, notice the large uncertainty spread of ±30% in the results. The magnitude of the uncertainty precludes any certainty in the real answer. Any experimental proof should be between 0% and 3%. Also, a major concern is the misalignment of the centroid points of each experiment. The non-centering can be displaced 10% favoring a smaller deflection. Examining the table again, shows that the best median value that intercepts the most experiment results is 1.78 arcsec. This agreement is within 3% of the predicted value, and should not be interpreted as the real final **GR** value. Why? Because there is no reported statistical distribution of the uncertainty. If the distribution is a uniform distribution, then the observational value can be -30%

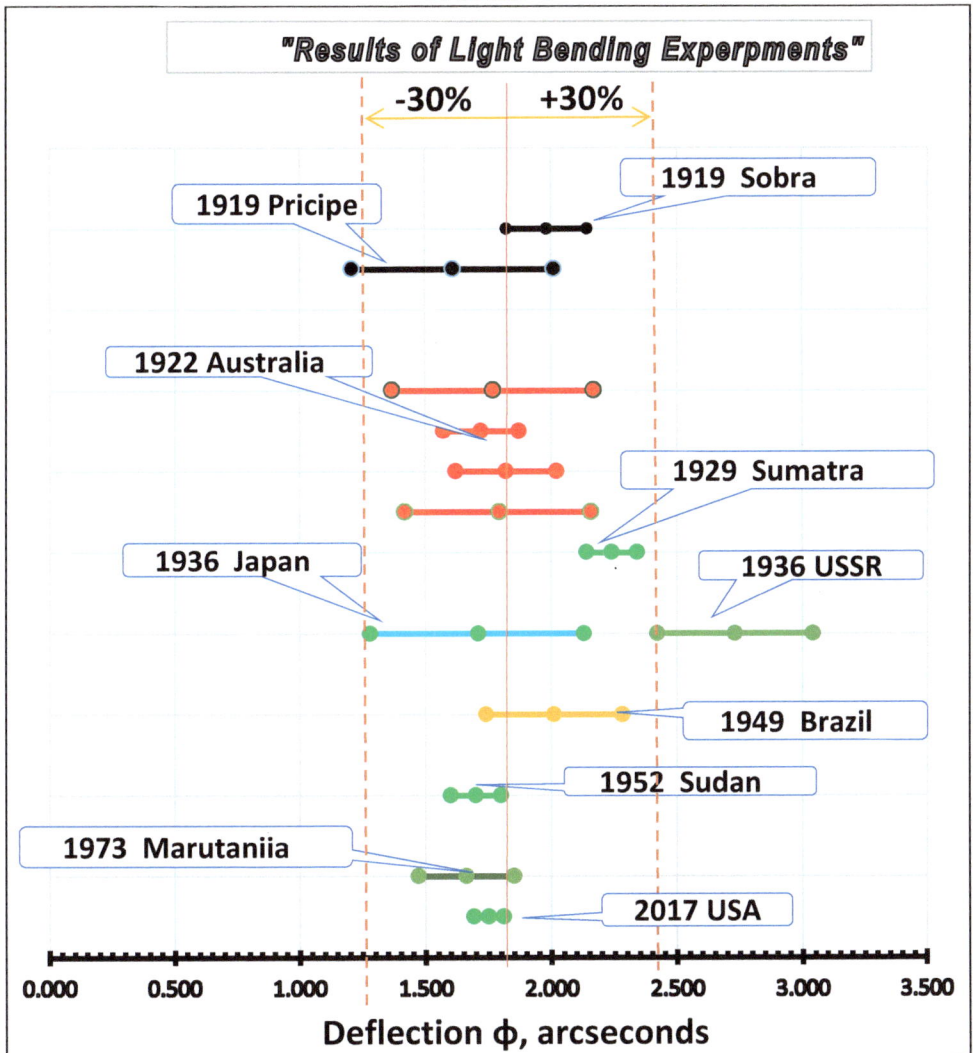

TABLE 1

in some cases. If it is a normal distribution, the value can be 1.78 arcsec with a 50% chance of being correct. Now, these numbers do not include the non-intercepted results for Sumatra and the U**SS**R. In this case, including Sumatra and the USSR, the centroid value would be 1.88 arcsec for a normal type distribution and ± 60% for a uniform distribution. As far as I can see, it is suspicious why results of Sumatra and USSR are not included. There is no explanation, that I can find, why these two eclipse observation opportunities are so different. If one chooses only the best results that have less than 15% uncertainty, the average median remains essentially the same. If we pick only the 2017 experiment, the uncertainty drops to less than ±3.5%. Some experiment error sources are of the same magnitude as the sought-after parameter. These include atmospheric turbulence, solar corona deflection, optics diffraction and distortion, equipment thermal and structure stability, star positions, camera exposure time, etc.

At this point and with the selectiveness we just went through, one can almost conclude that Einstein's **GR** has been proven correct. But, there are major misgivings reported by Group B described above. Two of these appear not to be included in the total model or data reduction process. The first is the diffraction of starlight around the Sun and moon, and the second is the refraction through the Sun's corona. Both of these bending effects can be depicted in the same manner as ***Illustration 16***. Now the deflection is strongly dependent on "miss" distance of the observed starlight ray from the the Sun and moon surface. Most of the early observations, for the early experiments, were taken at a solar miss distance of 3 to 8 solar radii, and most likely have very little impact on the findings. However, the 2017 observations were mostly taken at 0.5 to 4 solar radii, and these may be an additional error source. It is important to recognize that no observation line of sight grazed the surface of the sun because of the obvious brightness of the corona. All the data reduction had to be for Einstein's **GR** predictions at an observation number of solar radii above the Sun's photosphere. This means that the predicted value of 1.75 arcsec at the Sun's surface had to be extrapolated or inferred. However, the moon is a different story. There would be extremely small or no bending due to weak lunar gravity, or refraction. However, there may be significant bending due to diffraction because observations would have been taken during eclipse and extremely close to the lunar surface. It's difficult to stop searching and eliminating error sources when one arrives at the predicted value of a theory proving parameter in question.

Actually, there are a number of other misgivings, but we will refer to only one to exemplify the point of this discussion. The reference regarding an objection to this conclusion is:

"Accurate Solution to the Gravitational Bending of Starlight by a Massive Object", Fengyi Huang, Journal of Modern Physics, 2017, 8, 1894-1900

In this reference, **GR** is <u>**not used**</u> in any way to analytically arrive at the same **GR** value of 1.75 arcsec. *"Hmm-m, a coincidence?", a different theory or another demonstration that various unresolved error sources can validate any theory having a validation parameter of the same magnitude as the error source(s)?* With this caution now expressed, we can proceed to the next topic.

3. Gravitational Red Shift
In Einstein's **GR**, it is stated that clocks tick slower at a higher altitude when in a gravitational field than clocks at the surface. It has been shown, supposedly, that this **GR** statement has been validated via experimental results. These experiments included the "Pound-Rebka" terrestrial lab experiment of 1959, the "Pound–Snider" rocket launched experiment of 1965, and the current Global Positioning System **(GPS)** satellite operations. As before, these experimental conclusions suffer greatly when inquired of the "up-the-creek" Group B community. One such Group B community member is Ron Hatch who actually worked on the GPS program. I would heartily recommend that you review his many **"youtube"** videos, especially:

"Electric Universe", 2013 conference in Albuquerque Jan 3-6.

This is only one scientist of many that are detractors of **SR**, **GR** and the **Space-Time Continuum** concept. We cannot elaborate on details of his research and findings because it is beyond the scope of this book. However, this author will present some thoughts and conclusions that are beneficial to the understanding and visualization of this topic.

CAUTION! "That shifty Red Shift"

Before we begin, there are two concerns that must be kept in mind. First, is that many Relativists have presumed that the red shift of a radiating body is due to the inherent physics of that particular body; i.e. hotter or cooler bodies. Radiating "bodies" (or sources) can range from photons to massive suns. Second, and most important, is that they do not consider the radiating body to be a "moving" or translating source. This latter concern dictates that there is no Doppler effect of a moving source which is inherently non-factual. These concerns force the consideration that the gravitational gradient is the cause of a red shift in frequency when gaining altitude in a gravity field. Also, when descending a loss of potential energy is converted to a blue shift. Hence, the application of **GR** becomes a necessary operative agent. Now, we can look at this phenomenon from the view point of a *moving* radiation source. **Illustration 17, "The Red Shift"** shows, the problem at hand.

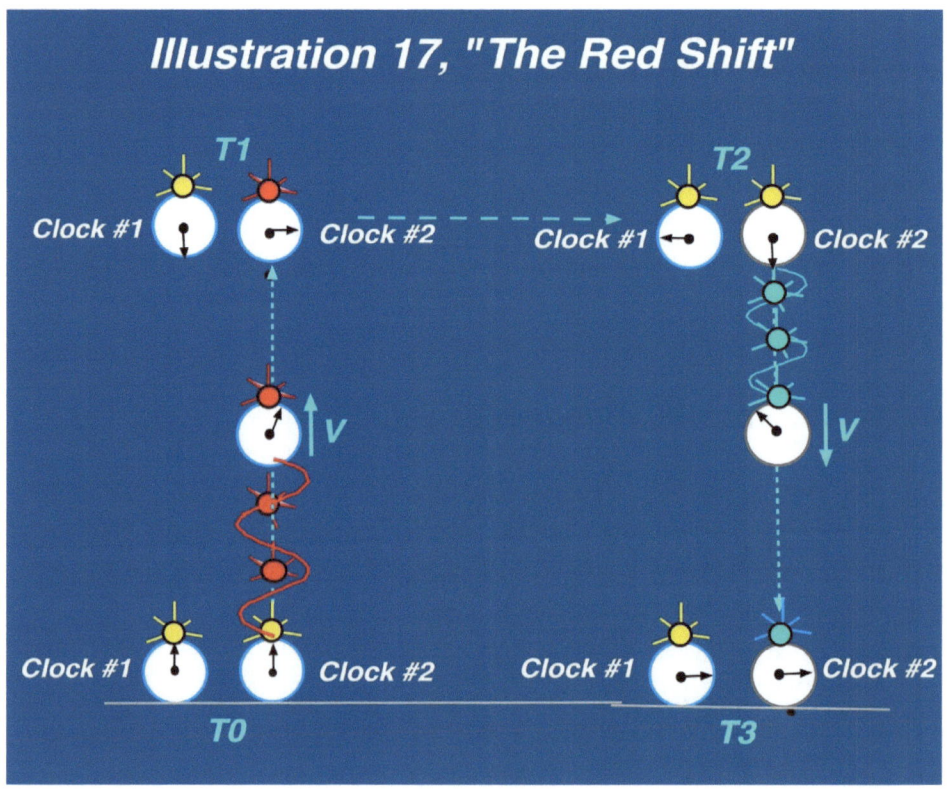

Beginning with the left side of this illustration, at time=T_0, we see two photon clocks. Both clocks are at rest on the "ground". Each clock is emitting a flashing yellow light with a wavelength **(λ)** of 570 nm (nanometers). They are both synchronized and in phase with their flashing emissions. and their initial times are reading 12:00 sharp. *Clock #1* remains on the ground at all times and is the observer's clock. The observer (not shown) is going to observe *Clock #2* as it translates or moves up and down relative to himself. In order to demonstrate some real numbers of what is taking place, it is assumed that **V=0.5c**, i.e. half the speed of light. We will now translate *Clock#2* at a constant velocity **V=0.5c** for one-half hour.

At T_1 =0.5hr, both clocks are shown for comparison of their relative time keeping. When *Clock #2* begins moving up, it becomes a moving radiating source and is receding from the observer. To the observer on the ground the clock is emitting a "stretched" or "redder" wavelength. This redder wavelength is now around 855 nm which results in *Clock #2* ticking 50% slower than *Clock#1*. Thus, the observed frequency has decreased by 50%. **Illustration 17** shows this comparison at T_1. *Clock #2* is interpreted to be running slower since it appears to be reading 12:15 rather than his clock that's showing a T_1 time 0f 12:30. Before we proceed, let's ask a question. "What if an observer is moving with *Clock #2*?" "What does he see at T_1?" Well he reads his *Clock #2* with a time of 12:30! Thus, we enter the world of **Relativity** where an apparent event is construed to be fact. Both observers can't be right in what *Clock #2* says, but one can be wrong. **Reference 1** in **Chapter 2** provides a fuller discussion on this dichotomy and resolves this confusion. Now at T_1, let *Clock #2* be brought to a halt from its upward movement

What happens to the frequency and emitted wavelength? The answer is that nothing happens! It is still flashing yellow consistent with a **λ= 570 nm** and synchronous with *Clock #1* except the apparent time of the observed *Clock #2* is 15 minutes slower. So, let's wait for a rest period of 15 minutes to observe both clocks from the ground at T_2. The question is; "Is the frequency change of *Clock #2*, while in transit, a permanent change as predicted by **GR** red shift theory?" Referring to the right side of **Illustration 17**, we see both clocks flashing yellow lights at identical frequencies. *Clock #1* shows the hand has advanced 15 minutes to 12:45. *Clock #2* shows the same 15 minutes advance to 12:30. This difference of 15 minutes in both clocks proves that the apparent frequency change during transit is not permanent or real, and that gravity and **GR** have no part in causing the red shift. The red shift is caused by the Doppler effect from a moving source. Now let *Clock #2* "fall" back to the ground. We now see a reverse process take place. Since *Clock #2* is now advancing toward the observer, the Doppler effect is a shortening of the wavelength to a 285 nm blue shift. Furthermore, *Clock #2* is now ticking 50% faster than *Clock #1*. Upon reaching the ground and stopping its descent, the *Clock #2* hand has advanced to a 3:00 position. and the *Clock #1* hand has advanced 30 minutes to the 3:00 position as well. At this time **T3**, the clocks are again synchronous and flashing yellow lights. The observed time bias lag during ascent is offset is by the observed time advance during descent. It is important to note that although both clocks maintain the same frequency throughout, that the Doppler effects must be accounted for in order to maintain optical or radio frequency communication links. This makes all the sense in the world since an atom's emission frequency is *unaffected by a gravity* field gradient. The subatomic nuclear binding force is infinitely greater than any disturbing force due to gravity!

This ends the discussion on this topic but it certainly shows that there are at least two sides to every story.

4. Gravity Probe-B (GP-B)
This experiment is probably the most expensive and technologically challenging experiment of all. This NASA program that sponsored this experiment was named "Gravity Probe-B (GP-B)" and cost $750M. This program also supposedly proved, after the fact, that the **Space-Time Continuum** concept was valid. Before we continue a more technical discussion, a brief history is warranted to set the stage.

GP-B satellite program was given the "go-ahead" funding in late 1963. The 21-foot-long, 3-ton satellite was launched into a 400-mile high polar orbit in April 2004, a period of 40 years from go-ahead with a data collection experiment life of about one year. The mission ended in 2005. Post-mission data analysis was performed with NASA funding through 2007. However, major and numerous error sources prevented any confident conclusions to be reached. Four more years would be needed to coax out **GR** parameters. Additional funding was obtained by the science members team from outside NASA sources. The objectives were to measure the "frame dragging" and geodetic curvature of **Space-Time** as predicted by **GR**. We are talking measurand rates of 6.6 arcsec (0.0018°) per year for geodetic effects, and 0.040 arcsec (0.00001°) per year for frame dragging.

Illustration 18, "GP-B Spacecraft Design", shows the in-orbit GP-B spacecraft rotating around the telescope optical axis one revolution every 78 seconds, in the hope of averaging or "zeroing-out" potential error sources. The four "windmill-like" solar arrays are pitched so that they receive sun generating power on every orbit revolution. The spacecraft was launched into a 90° polar orbit which means that the orbit is practically unaffected by Earth's oblateness and remains nearly fixed in inertial space. This in turn means that, all things being perfect, the spacecraft's telescope would be continually aligned to the line-of-sight (LOS) toward the guide star IM Pegasi. Further, IM Pegasi is located above the Sun's ecliptic plane, and thereby is visible to the spacecraft throughout the year. On each orbit revolution, IM Pegasi is occulted by the Earth and reacquiring this guide star each orbital pass can be a source of many errors that must be accounted.

The most important design feature are four independent spinning spherical gyros that are electrically suspended within their cryogenic cooled containment modules. The spacecraft is controlled in attitude and velocity to avoid any physical contact with the gyros. Precise control is needed with feed back from the gyro clearances to compensate for atmospheric drag affects, solar radiation pressure, outgassing and unequal tail-offs of vented "thrusters" of the control system, etc. In effect, the spinning gyros are "free-falling" in inertial space. In fact, this unperturbed free-fall is crucial to the conduct and success of the experiment. It is especially noteworthy that, the gyros are a mere 1.5 inches in diameter and were expected to spin at a rate between 7000 rpm to 10000 rpm. However, during the data collection phase of the mission, it appears that the actual spin rates were about 4000 rpm. Also, the spin rates decreased periodically because of residual helium gas molecules within the gyro modules generated by the spin maintenance operations and possibly by in-gassing from other sources. Since the gyros are in free fall, they can be visualized individually as separate entities as shown in **Illustration 18**. If all physical producing torques, misalignments, modelling uncertainties and the like acting on the gyro can be accounted for, then the resultant spin axis precession is the supposed true effect of curvature of the **Space-Time Continuum**.

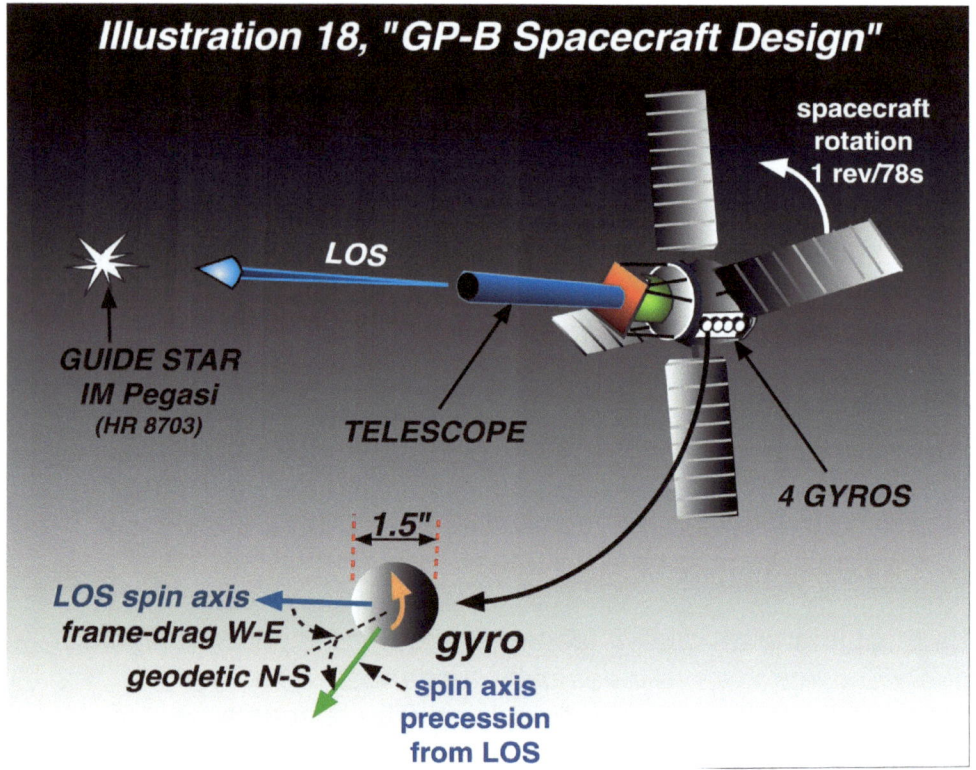

As depicted in **Illustration 18**, the precession of the gyro spin axis from the guide star LOS can be represented by two components. One component is a West-East component that determines frame drag of the **Continuum** due the Earth's rotation. The second North-South component determines the curvature or geodetic effect of the **Continuum**. And therein lies the overwhelming challenges to the design, manufacture, modelling, error source identification, operations and data processing necessary to implement the GP-B program. ***Illustration 19, "Space-Time Frame-Drag and Curvature"***, defines these two components of the spin axis precession. In ***Illustration 19, A*** (top depiction), shows a free-fall gyro in polar orbit with its spin axis precessing from West to East. The **Continuum** frame, looking down from the North Pole, is shown as being twisted and "dragged" because of the West-East rotation of the Earth. The amount of West-East precession from the guide star LOS is the the frame-drag effect.

Illustration 19, "Space-Time Frame-Drag & Curvature"

Illustration 19, B (bottom depiction), shows the gyro in orbit as seen from the equator where the **Continuum** curvature is very evident. Now, the amount North-South precession from the guide star LOS is the **Continuum** curvature effect. Notice that in *Illustration 19*, two Phantom gyros, with question marks inside, are displayed. It is also remarked that two of the gyros had their spin axes initially aligned (+) along the guide star LOS, and the other two were oppositely aligned (-) to the LOS. We will come back to these questions a little later. But first, I would like to give the reader some idea of the typical error sources and compensations that had to be addressed during the GP-B program.

Error sources: Beside the goal to design and manufacture near-perfect spherical homogeneous 1.5 in. gyros, the GP-B spacecraft involved many such design challenges, too numerous to mention here. But the greatest challenge of all was to identify and model error sources that might mask the **GR** parameters of frame drag and geodetic effects buried deep in the collected data. This task is referred as data processing which was performed mostly after the GP-B mission ended in 2005. The average amount of "snap-shot" data collection was 216,000 points (as far as I can tell) for each gyro. Thus, began a complex sleuthing effort to find and model all contributing error sources so their effects could be eliminated from the data. There were over 100(?) error sources accounted for, some of which are listed below to give the reader an idea of their types. It is noted that the magnitude of many of these errors were equal to or even greater than the **GR** precession verification parameters, namely; N-S geodetic effect, -6606 mas/yr (1.835°), and E-W frame drag, -39.2 mas/yr (0.0109°), ("mas" = milli-arc-seconds).

Partial error source list

- guide star telescope alignment & realignment after each orbit occultation
- proper motion of guide star position
- gyro drift rate & spin maintenance operations
- rotating spacecraft induced torques
- 6-dimentional attitude control induced torques
- misalignments caused by thermal distortions
- spacecraft orbital motion & Earth gravity model

- gyro spin axis readout accuracy & control
- fixed and moving dipoles
- electric charges
- data processing algorithms
- yet unknown error sources
- etc.

At this point, let's take a look at the final results of the GP-B experiment as published by the American Physical Society in its open Physical Review Letters, PRL 106,221101 (2011). The authors are from the science team of Stanford University, Marshall Space Flight Center (NASA), and King Abdulaziz City for Science and Technology. Oddly, there is no evident joint publication or approvals by NASA Head Quarters, as you would have expected. **Illustration 20, "GP-B Final Results-Dec 2010"**, shows the final results of the four gyros and their resulting **GR** geodetic and frame dragging parameters. All the gyro two-parameter error ellipses were "combined" to show a much smaller error ellipse with its center at the **GR** prediction values. Thus, allegedly proving that Einstein's **Space-Time Continuum** concept is correct and therefore real!

<u>CAUTION!</u> 🚩 "The rest of the story"

Before we discuss these final results further, it is important to set the stage for waving this caution flag. The reader must be aware of the rest of the story that raises skepticism, serious concerns, and problematic questions in this author's mind. First, let's begin with the GP-B experiment program initially funded by NASA with a program cost of $750M. I've already mentioned the oddity that, except for minor press releases, very little credit or kudos are afforded or "claimed" by NASA HQ, nor does NASA present any, that I can find, significant or oversight publications supporting and approving the findings and conduct of the program. However, after a program review, NASA ended the GP-B contract funding in September 2008. The reason being that the collected data were too "noisy" and that there were other programs that had higher funding priority. Despite the Stanford University's science team's optimism and recommendation for NASA to grant more funding to continue data processing toward completion, the NAS Science Advisory Committee (SAC) declined to extend funding for three to four more years. But the science team persisted and found additional funding of about $2M to $3M from outside sources to support a dozen personnel for several more years. It should be noted that some members of the science team spent nearly 40 years, an entire career, on this one program and continually forecasted a successful completion. It seems very peculiar that NASA would not grant this request since this would only amount to $1M per year out of a $7B per year budget and, if successful, could result in great accolades. Evidently, they lost faith in any successful meaningful outcome and decided to cut their losses or play it safe (my opinionated conclusion). In any event, the program became a low priority item for NASA HQ.

A second aspect of this story is the fact that no one person or group could ever prove or disprove the concluded results and findings of this orbiting experiment. Why? Because no one could ever assemble a new team with the same experience level as the 40-year team that built, launched and operated the GP-B spacecraft.

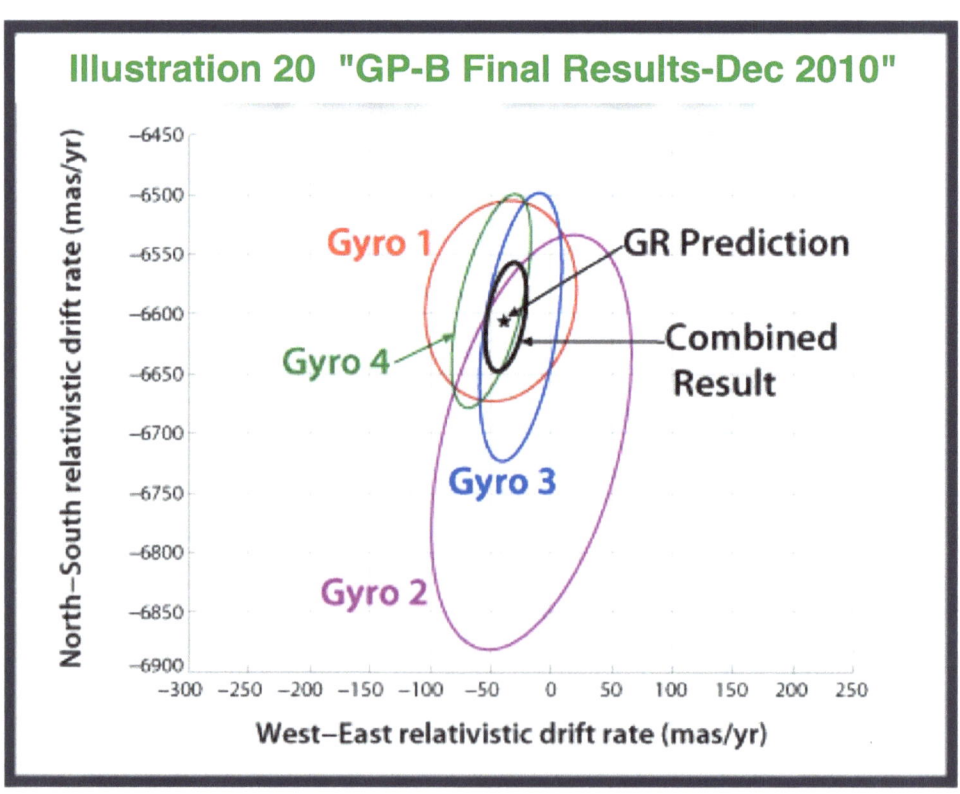

Further, there would be literally very few previous engineers and scientists to call on to answer the millions of questions that would arise. Besides, nobody or organization in their right mind would incur the cost. In fact, another disappointing NASA decision that happened in the same time period, was the cancellation of the proposed 2012 launch of the Satellite Test of the **Equivalence Principle, "STEP"**. This follow-on proposed program would be designed to test the **"Equivalence Principle"** of Einstein's **GR**. It too would involve using spacecraft with different mass types in free-fall. However, it should be no surprise that NASA would make this decision based on their posture relative to the GP-B program. Perhaps, they felt that any science returns would be "noisy" as well and would not justify the program cost or be credible.

Statistics story

Now let's return to ***Illustration 20***. It's amazing that this $750M 40-year program boils down to one simple sheet of paper of statistical results. This Illustration shows the data processing results of the four gyros in terms of 2D "2σ" error ellipses. Whenever statistics are used to prove anything, this author's eyebrows raise. The question of repeatability and robustness enters. What do "2σ" error ellipses mean? Well first, an error ellipse is simply the dispersion or spread of all the collected processing data points for one given parameter (1 Dimensional), two parameters (2D), three parameters (3D), etc.0 The parameter(s) in question is the unknown variable(s) you are seeking to solve. The error ellipse size depends on the density and distribution of the data points. For example, a normal independently random distribution would have a greater density near the center of the ellipse and less and less density as you move away from the center. This distribution type is what was applied in ***Illustration 20***. At first glance, it appears that a very confident solution substantiating the **GR** parameters was arrived with the "combined result". Now "2σ", pronounced "two-sigma", is a statistical mathematical term meaning that 95% of the collected data lie within the ellipse. Or in other words, the confidence that the correct solution is within the ellipse is 95%. Similarly, a "3σ" ellipse implies a 99.7% confidence that the correct answer lies within the ellipse, and a "1σ" ellipse implies a 68% confidence level. It follows that the ellipse size for a "3σ" ellipse is three times the size of a

"1σ" ellipse, and a "2σ" is two times the size of a "1σ" ellipse. Why are the various "sigma" values important? Because they can be influenced easily by arbitrary selection or *de-selection* of the considered data points. For example, there can be present certain "outlier" data points which can be preferentially cleaned or "tossed out" of the data point data base, thereby reducing the size and corresponding "sigma" of an ellipse. In addition, specifying a different arbitrary confidence level, say 98% instead of 95%, would increase the size of the error ellipses in **Illustration 20** by 20%. Then too, any departure from a standard normal distribution can significantly adversely affect the sizes of the ellipses. So, there can be serious concerns and skepticism about the statistical portion of processing, especially when knowing what solution you are driving to; and the easy manipulation that can be applied. In fact, the "combined results" ellipse shown in **Illustration 20** can be misleading. It is not a "2σ" ellipse because, as it will be proven below, it is a uniform distribution which means the error ellipse has, at best, only a 50% confidence level. What we really see in the combined results, viewed from set theory, is that this combined ellipse contains all data points common to all the gyro ellipses at the "2σ" level. Does it prove the sought-after validity of the **GR** parameters? Yes and no depending on your argumentative interpretations. We can stop right here with our "no" answer as we read on further.

There are other aspects that should be considered, and that is repeatability and simultaneity. The GP-B science data processing turned out to be extremely sophisticated and complicated that Ph.D. theses could be written on each level of data processing. Every effort was made to produce only "good" data to be included in the final data process. Only "smooth" and contiguous data were included in the data base collection time segments. Erratic data due to acquiring and re-acquiring the guide star, solar flare interruptions, spacecraft anomalies, and the like were discarded. On an average, each gyro collected 213 days of science data. Data was NOT collected or used when the guide star was occulted by the Earth during each orbital revolution. As far as I can surmise, data points consisted of sequential sampling (40 second snapshots) for a total mission data base of about 216,000 data points for each gyro.

Now, if we take a closer look at this "combined results" ellipse, we find that only between 2% and 5% of the total data points are contained in this ellipse. From an engineering perspective, this repeatability is a far cry from desired. This infers that either the gyro error sources were not all accounted for, not sufficiently modeled, or gyro design manufacture and operational performance was inadequate (not robust enough) for this experiment. This inadequacy is confirmed because the centers, or expected values of the gyro ellipses, are too far apart from the **GR** solution. They should be more coincident with each other and have very much less "2σ" dispersions, especially gyro 2. However, it must be said that a N-S precession effect appears definitely present. With this comment we must go back to the question marks **("?")** shown in **Illustration 19, *"Space-Time Frame Drag and Curvature"***, and the **Space-Time Continuum** *Flaw #1* pointed out in **Chapter 1**. First, as a reminder, *Flaw #1* stated that the **"Continuum"** theory is derived utilizing fluid dynamics analytics around a *"closed body"* as the basis for the **"Continuum"** concept. In other words, the **Space-Time Continuum** cannot exist within the solid body. But, a spacecraft is a type of solid body with the gyro payload encapsulated within. If the theory is correct, then the **Continuum** "gravity" effects should be zero where the gyros are located, or the theory must be modified and extended to have a degree of porosity with an internal gravity field. But porosity, I don't believe is the answer since other spacecraft have "landed" on small asteroids under attraction of the asteroid's own miniscule gravity. Now in **Illustration 19-**A, we see that for a N-S pass, frame dragging effect (if it exists) on the gyro spin axis precession is from W-E direction. However, on the S-N pass of the orbit, when the guide star is occulted, frame dragging is still W-E but the amount of precession, (now westward of the spin axis) from the N-S pass is being cancelled. This cancelling effect is also present for N-S geodetic effects. Note again, that no data sampling was used during guide star occultations which lasted about one half an orbit revolution. So, the two concerns that have to be addressed are:

How can the gyros sense the **Space-Time Continuum** where there is supposed to be none inside the spacecraft, and why isn't a cancellation or "zeroing-out" process been detected if a **Continuum** is present?? Cancellations may occur at least four times each orbit, two for frame drag and two for geodetic precession. The answer may lie, in part, with the unexpected jumps of 100 mas in precession encountered during each orbit revolution or how the data was processed on only the N-S pass of the orbit. These numerous unexpected jumps have been attributed to resonant conditions between the spacecraft constant roll and the gyro spin control. The jumps may have nothing to do with **GR** or roll resonance but simply an "unloading" of the cumulative applied torques from an opposing <u>unknown</u> torque, sensed by the spin control subsystem.

<u>CAUTION!</u> *Statistical False-Positives*

Now, we are going to take a step backward in the above processing scenario. The following discussion is going to be somewhat convoluted or difficult to follow, but really very important on putting a measure of value to the final gyro statistical results of ***Illustration 20***. What we will be doing is to connect the statistics of a previous processing step *(Illustration 21)* to the final statistical results in ***Illustration 20***. You will be confronted by a very serious and grave apprehension in the "connection". The connection step is shown in ***Illustration 21 "Gyro 2 Processing Results"***.

Illustration 21 presents the processing results for only Gyro 2. There were 10 segments of science data collected for each gyro. However, it seems that two segments were too short and two were not of sufficient quality to include in the data base. Therefore, only 6 segments included are as shown in ***Illustration 21***; and at first glance, it's unbelievable! You would think that these were data from six different gyros rather than from just one. There is no correlative evidence that the results are from a single gyro. The error ellipses are literally off the chart ranging from -2000 mas/yr to -12000 mas/yr for geodetic effects(N-S) and -2000 to +4000(W-E) mas/yr for frame dragging effects. These data spreads are ±100%(N-S) and over 2000%(W-E), completely out-of-sight of **GR** predictions of -6600 mas/yr(N-S) and +40 mas(W-E). This kind of outcome would be immediate cause for failure and unacceptability. No wonder NASA lost confidence in obtaining more "concrete" evidence.

Continuing on, it appears that the science team's reasoning was to determine the overlap of the 2σ ellipses and, lo and behold, the "set theory" overlap luckily contained the predicted **GR** parameters. This overlap is pointed out as the "combined ellipse" in ***Illustration 21*** which represents the Gyro 2 statistical 2σ ellipse in ***Illustration 20***. But the combined ellipse in ***Ilustration 20*** is really representative of a uniform distribution which implies only a confidence level of, at best, 50% and not 95%. Actually, the confidence level of the combined ellipse is more likely 5%. Therefore, the "2σ" ellipses shown in ***Illustration 20*** for all the gyros are in error.

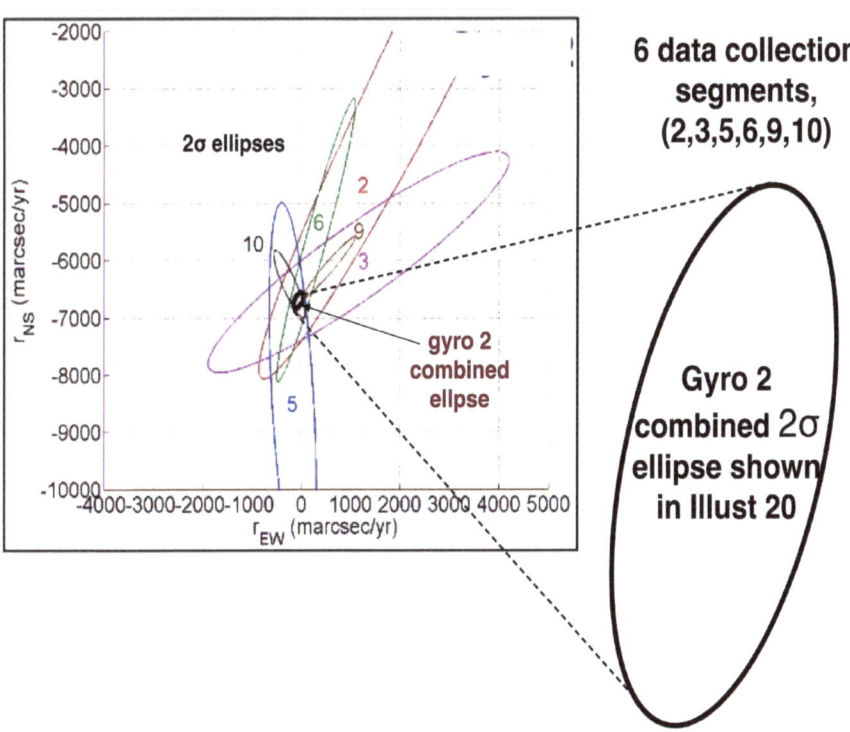

Illustration 21, "Gyro 2 Processing Results"

If we study the results of ***Illustration 21*** further, it is apparent that six selected segments of data are an insufficient number to characterize the gyro. It is also apparent that there is little correlation between the various segments, hence the appearance of being that of separate gyros. The same gyro should have had more correlation in that their ellipse major axes should be aligned probably within 5° or so. Also, the magnitude of the major and minor axes should be approximately the same and within a spread of a few %. This type of processing quality output would have given "robust" confidence that the gyro is performing well and that all the error sources, modeling, and processing algorithms are acceptable and accounted. Each gyro would have a similar, but individualistic signature of its own. But, such is not the case! Gyros 1, 3, and 4 all exhibit similar unbelievable traits as Gyro 2.

So, what can we surmise about all this? First, the combined Gyro 2 ellipse in **Illustration 21** does NOT represent a normal distribution and when expressed as such in **Illustration 20** is a false-positive. It really, more likely, represents a uniform distribution; implying that at any point in the so called 2σ ellipse there is only at best a 50-50 chance that it is the correct solution and a more probable chance of 5%.

Second, the number of data points within the overlap or combined ellipse **(illustration 21)** area is probably less than 2%, or less than 4000 data points out of 216,000 points collected. This is woefully a small number indeed and does not support any measure of repeatability or simultaneity! Also, a false-positive.

Third, what this in turn means is that less than 600 points are contained in the combined ellipse shown in **Illustration 20** with a confidence level of less than 5%. Another **Illustration 20** false-positive.

It is extremely difficult to accept the lack of contiguous non-random processing solutions. For example, two sequential 40 second snapshot data points produce processing estimate differences of 3000 to 5000 mas. There is absolutely NO simultaneity or robustness attributes present! These attributes should be expected in a normal smooth operating satellite over an 80 second period. Completely unexpected and suspicious! Results are nowhere consistent with a robust experiment. No wonder NASA said the results are too "noisy" to be credibly accepted.

This author strongly feels that it is incorrect to represent each processed segment as a different observation source. It is not! They are from measurements of the same gyro with the same type of observations, measurands, drift/spin controls, data readout, error modeling, and data processing technique. I would simply pose this question to the science team, "What if the data collected for the six segments were completely contiguous (no interruptions)?". Would you not process them as being that measured from a single observational source (the same gyro) and not in separate observation segments?

Now, the "case" for using set theory to determine the common overlap is certainly unbelievable! I believe that the science team faced the problem of insufficient number of data segments and really may have required twenty to thirty more segments and therefore used an illogical set theory approach as an artifice to encircle the predicted **GR** parameters. However, I don't believe a longer mission would have helped anyway, because of the continuing far ranging spreads and random ellipse orientations of the resulting processing. The author calculated a "guesstimate" assuming a contiguous accumulation of more segments with the same type of data spreads. The guesstimate indicated a mean value of -6040 mas(N-S) and +630 mas(E-W) with a 2σ value of 4800 mas and with a near zero correlation between the N-S and the W-E **GR** parameters. It makes you wonder why the other two collected data segments were not used. Apparently, there was a lot of data scatter that didn't fit, and the scatter couldn't produce a set of ellipses to contain the predicted **GR** parameters. Again, the experiment suffers from a serious lack of robustness. It wouldn't surprise me if the omitted segments actually supported the author's single "guesstimated" error ellipse. This leads one to question whether another unaccounted physics mechanism, error source or data processing modelling error is producing this "near" **GR** value of -6040 mas precession. This author's opinion is that this result, even if only approximately correct, is completely unacceptable as proof of the **Space-Time Continuum** concept. The final GP-B experiment results, as presented in ***Illustrations 20***, are completely misleading and errored!

Listed below, in no particular order, are a few other concerns and questions that confront an interested outside observer and further cast a doubtful dark shadow on the results.

• If the gyros do "feel" a **Space-Time Continuum**–generated torque, then there must be an energy loss in the Earth's inertial orbital speed and rotational speed; – also from a mesh compression wave drag effect. Error source and inadequate modeling?

• Is JPL star and planetary ephemerides, and hence proper motion of guide star, already corrected for **GR** effects(?). May be source of error because the **GR** theory is still in question.

• Differential gravity acting across 1.5" diameter gyros and spacecraft structure may have a disturbing torque on the order 10^{-7}, which is the same magnitude as other error sources. Another source of error?

• Attitude and Translation Control (ATC) subsystem slaved to a separate non-spinning free-fall mass and not optimized to every individual gyro to maintain more exacting free-fall conditions on each gyro. Error source?

• 3-D mesh simulation of **Space-Time Continuum** (see **Chapter 1**) appears not to be fully considered in data processing algorithms. Modeling error?

• The Earth is "nested" in the Sun's dominant **Space-Time Continuum**— but the Sun's **Continuum** is not flat at location of Earth. Error source and modelling error?

• Seasonal effect of sap rising & falling in trees on Earth's moment of inertia and rotational rate. Modeling error? (*just being facetious here . . maybe*)

This concludes our discussion of **Chapter 5**. There are too many open questions and concerns to accept any conclusions of this experiment. We are now ready to proceed to the most perplexing **SR** paradox of all; the famous "Twin" Paradox.

Chapter 6

"The twin paradox: What is it?"

Perhaps the most difficult **SR** paradox to accept or intuitively understand is the twin paradox. As alluded to in the **Introduction**, there are two 20-year-old living twins on Earth. One of the twins departs and returns on a ten-year round-trip space voyage traveling at close to the speed of light *(0.99c)*. When he returns he finds his brother (or sister) who remained on Earth has aged and is now 90 years old while he is only 30 years old. How can this be? It's just not fathomable! It's all part of **Einstein's Special Relativity (SR)** theory. **Reference 1** (referred in **Chapter 2**) has already published the serious flaws in **SR** and **Chapter 3** of this book has restated and elaborated upon these flaws. How do we explain this?

Riding A Light Beam
Let's jump right in and show **Illustration 22 *"Riding A Light Beam"*** to begin to understand and interpret this paradox. This paradox was posed when Einstein, at a young age, imagined himself running along side a light beam and was puzzled by what he visualized. It was later in life when he revisited this imagination and solidified the interpretation and its meaning along with firming his **SR** theory and the related consequence of **time dilation**.

Looking at **Illustration 22**, Einstein was riding a train and looking back at the town clock and visualized himself riding or running faster and faster along side of the light beam emanating from the clock. At **T=0**, both the town clock and his wrist (or pocket) watch were synchronized at 2:30. First, he began running at half the speed of light *(0.5c)*. At 15 minutes elapsed time, he looked at the light beam image along side of him and observed that the town clock image read 2:45, but his wrist watch showed a time of 3:00. He repeated his marathon, but this time he ran at the speed of light *(c)*. Again, 1 hour later, he looked at the town clock image and he was amazed that the town clock was stopped and still read 2:30. But his watch read 3:30, an elapsed time of 1 hour. He was puzzled by these observations but reconciled them by concluding that anyone running along side the light beam (an outside observer) would have a wrist watch running faster than the clock image seen by a local moving observer in the light beam.

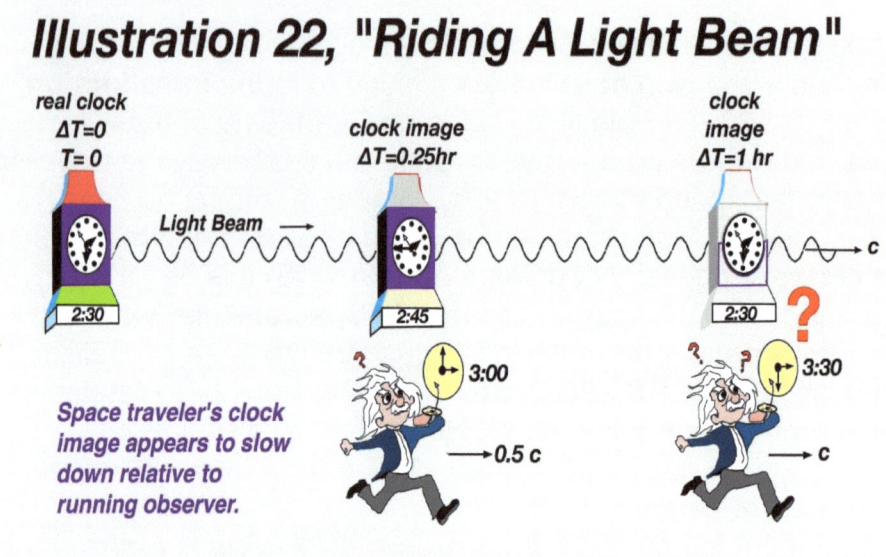

Thus, the Earth-bound twin (the runner) ages faster than the space traveling twin. This is the essence of the twins' paradox. Einstein was baffled by the visualization of a "frozen" light wave when running at the speed of light as shown previously by **Illustration 11** in **Chapter 3**. He did not visualize the clock was losing its color and beginning to fade. Einstein did not allow for any doppler effect (**Flaw #3, Chapter 3**) as observed by a moving space traveler. He was convinced that such a "frozen" condition could not exist. Thus, by further postulating that a moving observer must "see" an oscillating wave, and therefore, must measure light speed as **c**. The consequences of all these visualizations and presumptions produced the famous **SR** singularity that **c** cannot be physically exceeded. In addition, peculiarities that observed lengths of objects shrink and moving clocks slow down became part of the **SR** development. In the following discussion, we will show that what he thought he observed was a perfectly normal phenomenon and fully explainable.

ALERT! *Flaw #7, "Don't blink you'll miss a flaw"*

Misunderstanding the frozen wave phenomenon is a major flaw. It is better to rephrase and refer to this phenomenon as a fade-out or/and "Blink-Out" phenomenon. ***Illustration 23, "The Blink-Out Phenomenon"*** demonstrates how the town clock image simply fades and finally blinks out as the speed reaches the speed of light. The fading is caused by the Doppler increase in wavelength with increase in speed **(V)** as experienced by a receding observer within the light beam. The outside running observer, if he could see, he would see the frozen wave at **V=c**. The wavelength of a particular color is "stretched" toward the infra red and then to a non-detectable optical wavelength. Since "red' is the color with the longest wavelength of 650 nm, it is the first to blink out. Note that ***Illustration 23***, for simplicity, intentionally omitted any time dilation or length shrinkage effects.

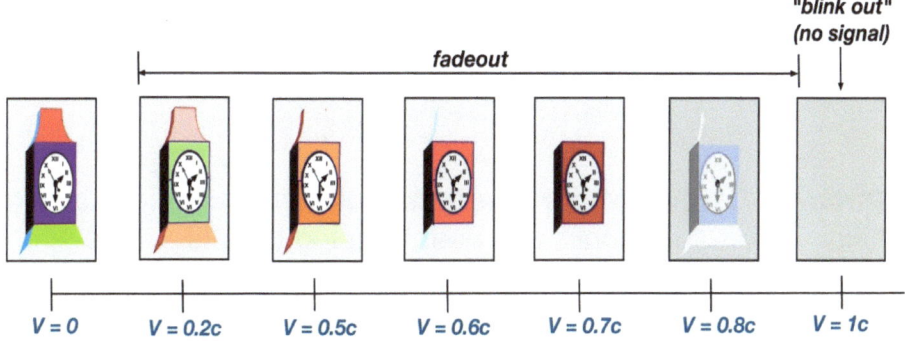

Illustration 23, "The Blink-Out Phenomenon"

red, 650nm @ V=.2c
green, 510nm @ V=.5c
blue, 450 nm @ V=.6c
violet, 380nm @ V=.8c

You might ask, "What happens if the running observer runs faster than the speed of light?". Well, I'm glad you asked that question. There is no reason why he can't run faster since there are no bounds to our imagination. Everything is just a figment of our imagination and we can imagine anything. The challenge is to avoid being deceived by our imaginations.

Passing A Light Beam

Actually, Einstein stopped short when he reached the speed of light *(c)* in his visualization. If he would have continued running faster, he would have passed the blink-out phase and would have again begun to see the image of the town clock. ***Illustration 24, "Passing a Light Beam"*** depicts what happens if he ran faster.

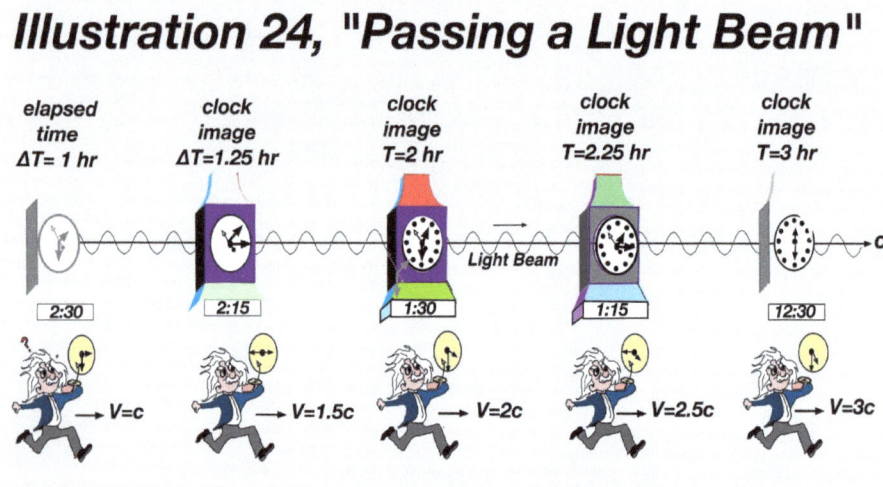

Space traveler's clock image appears to reverse relative to running observer.

Illustration 24 begins where ***Illustration 22*** left off; namely, at ***V=c***. Beginning at left, the runner has been running for 1 hr. at the speed of light. His wrist watch reads 3:30 and the clock image in the beam reads 2:30 as he exits the blink-out period. Now if he was running at a speed of ***1.5c*** for 15 minutes (0.25 hr.), His wristwatch would read 2:45 and the town clock image would show 2:15. Hey what's going on here? Let's do this again, except, let him run at a speed of ***2c*** from exiting blink-out for 1 hr. Now, his wrist watch reads 4:30 and the clock image shows 1:30. Sure enough, the time in the clock image is actually going backward. How can this be? How do we interpret this backward movement? Simply! We are now looking into the past. What has happened is that we have caught up to the emanated light beam that was emitted at **ΔT= -1hr** *(refer to **Illustration 22**)*. Also, at this time, the true colors of the clock are again visible. As the runner speed increases to three times the speed of light ***(V~3c)***, the image colors tend toward the high frequency violet range due to Doppler effects. At ***3c***, they enter a non-visible black-out region. The runner runs for two hours and his wrist watch now reads 5:30 and the clock image shows 12:30. Why is the above discussion important. Because it will give the reader a better intuitive and common sense understanding of the twin paradox involving a manned round-trip interstellar trip to a nearby star. And now, let's prepare for our interstellar round trip.

Chapter 7

"The twin paradox debunked"

All Aboard for round trip to Alpha Centauri

As part of our preparation, it's very desirable to understand the aspect of space communications. Communications consist of radio or optical transmissions originating from earth and being detected and received by a spacecraft or starship and vice versa. Although communication principles are widely known, we include the basics herein for the reader's convenience. **Illustration 25, "The Communication Basics"** presents a typical communications example.

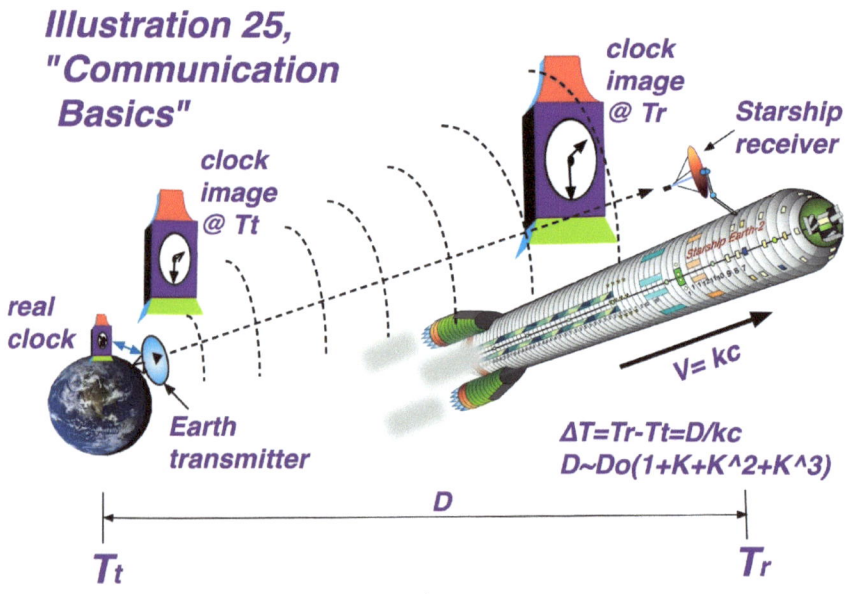

The Illustration shows a real physical town clock on the Earth. A digital snapshot is taken with 2:30 showing on the clock face. The digital image is then transmitted at **Tt** immediately to the Starship via an Earth based antenna. At the same identical time **(Tt)**, the Starship transmits a message to Earth. Let's say that the Starship is one tenth light year (0.1 Ltyr) away from Earth, which is the equivalent distance **(Do)** of 5.88E+11 (588,000,000,000) statute miles. Further assume that the Starship is traveling at a speed **V** of **0.35c** and moving away from Earth. Note that **k=V/c=0.35**. Now, the propagation time of the transmission to reach a distance of **Do** is 36.5 days. But the Starship is not there but has moved on during the time it took to propagate the signal. When the signal is finally received **(Tr)** by the Starship, the elapsed time from **Tt** is 76 days, and the distance **(D)** is 0.14 light years from Earth. Note that the digital image received shows its true colors as seen on Earth. However, the signal transmitted by the Starship at **Tr** arrives at Earth **Tt +** 36.5 days. Note that the Starship signal is not shown in the Illustration. The point of this discussion is that when the signal is received by either party, there is a delay between its transmission and reception. Any image that is transmitted is an image from the past. For example, the clock image received by the Starship in **Illustration 25** shows 2:30 which is the image taken at the time of its transmission. But the real clock on Earth may show 2:30 but will be dated 76 days later than **Tt**. When the image transmitted at **Tt** is received by the Starship, it is received at **Tr=Tt +76 days**. Further, a digital relayed image is NOT affected by Doppler, but an optical image received at the Starship (e.g. through a telescope) will be affected by Doppler.

Now, one might wonder, what a space traveler might view outside the Starship while he is in motion. **Illustration 26 "A View from a Starship"** depicts such views when looking aft and looking forward. This view will add a further insight into communication operations. Pretend for a moment to be traveling in a Starship moving away from Earth at a speed of **V=0.35c**. You walk to an observation viewing dome at the stern of the Starship and look at the surrounding star field. You notice that the stars have a rainbow of hues ranging from yellow (at the outer edge) to an increasing reddish hue directly in the middle of the starfield. The rainbow effect is due to Doppler caused by the Starship's receding speed. The light wave lengths emanating from the stars are increasing as detected by the speeding Starship and thereby are "red-shifted". In **Illustration 26**, a +90° viewing angle is directly overhead and -90° is straight down. Looking at the extreme left and right edges are representative of looks to the right or left of the view.

If we were receiving a laser type signal from earth there would be a problem with reception. The signal received would be exhibiting the

Illustration 26, "A View From A Starship, 0.35c"

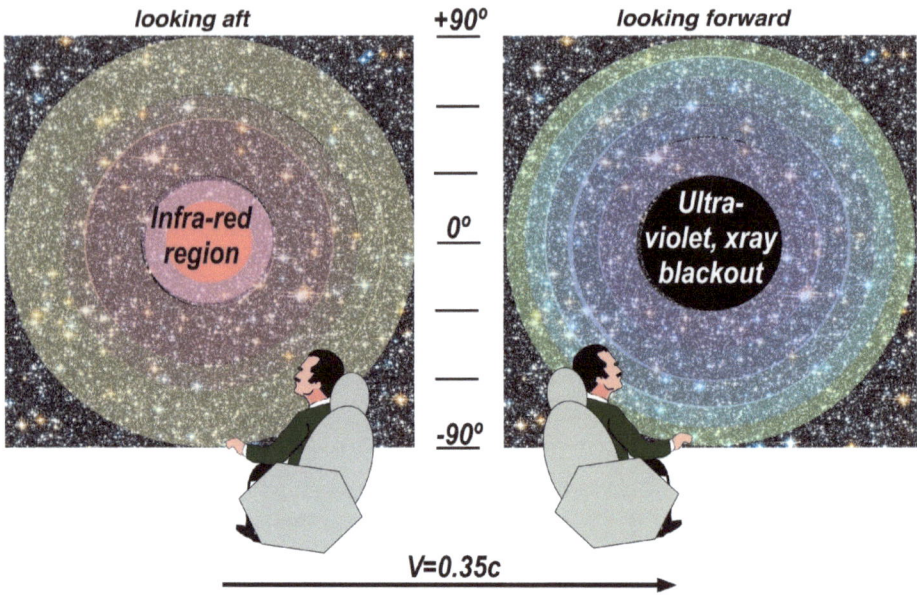

same red shift as the middle of the severe infra-red zone shown in **Illustration 26**. If the Starship also used a laser array, Earth reception would have a similar reception problem because of the affect of wavelength increase from a moving source; namely the Starship. If communications were conducted using radio frequencies with much shorter wavelengths, the communications would suffer even a more severe problem of detecting and maintaining radio links.

Now, let's go to the bow of the Starship and take in a forward-looking view of the visible starfield. We immediately notice in **Illustration 26** a similar rainbow of colors but in a range from green to deep indigo. However, there is one very noticeable feature and that is; there is a blank empty circular "hole" in the middle of our view. This second half of the rainbow colors is due to Doppler shortening of wavelengths when the Starship is advancing toward the starfield. In this case **(V=0.35c)**, the observed wavelengths reach the nonvisible range of ultra violet and xrays. Again, communication problems are not the same as looking aft but they are of the same order of difficulty. "Are we ready to embark on our interstellar trip?" "Not yet!"

Round Trip Objective
Before we depart, we must be reminded that the objective of the trip is to test the theory of Special Relativity **(SR)** and its corollary famous "twin paradox". The manned round way trip will be at a speed of **0.95c.** In order *to* understand what we will expect to experience at this high speed, it was imperative to first understand the previous basics of communications. Now, we have to be ready to encounter the unexpected; that is **SR** peculiarities vs. the well known and currently implemented communication truths in our every day lives. ***Illustration 27, "SR vs Communication Truths"*** accomplishes exactly this awareness.

In discussing this ***Illustration 27***, a little patience is necessarily required. First, three different observation viewpoints (**A, B, C**) are presented. '**A**' observation is what is being observed aboard the Starship in real time. '**B**' is also a real time **SR** direct observation (if he could actually see) by a person on Earth. '**C**' is also a real time **SR** direct observation by a person on Alpha Centauri 4 light years from Earth. '**B**' and '**C**' observations are reversed for return trip. Let's look at the two observers and what they see as supposed "truths" of **SR** theory. The Earth observer "sees" the on-board traveler as being contracted and compressed. **SR** predicts that the contraction will be 30% at a receding speed of **0.95c.** On the other hand, the observer on Alpha Centauri "sees" an extension of the on-board traveler of 300%. This contraction or extension applies to the Starship as well.

Note that the contraction and extensions are in length only and not in height or width. It is mentioned here as a side tangent, that **SR** specifies that a length contraction occurs whether the Starship is receding from Earth or advancing toward Earth. Although this condition makes no sense, Einstein justifies it by stating that over a long unrealistic looping arc, the **SR** equations "hold". I'm not sure about this, so we will retain a "stretch" as shown in the Illustration. It won't matter in the end anyway as you will see in the final analysis. Another extremely important observation is that every one "sees" the town clock in its true colors. There are no Doppler effects! In referring to **Chapter 3**, Einstein forces the on-board traveler to see the same physics as the Earth observer (and vice versa), and hence NO Doppler. Thus, **SR** must compensate for this (Postulation 2, **Chapter**

2) by contracting lengths (in the direction of flight only) and dilation (expansion) of time.

Illustration 27, "Special Relativity vs Communication Truths"

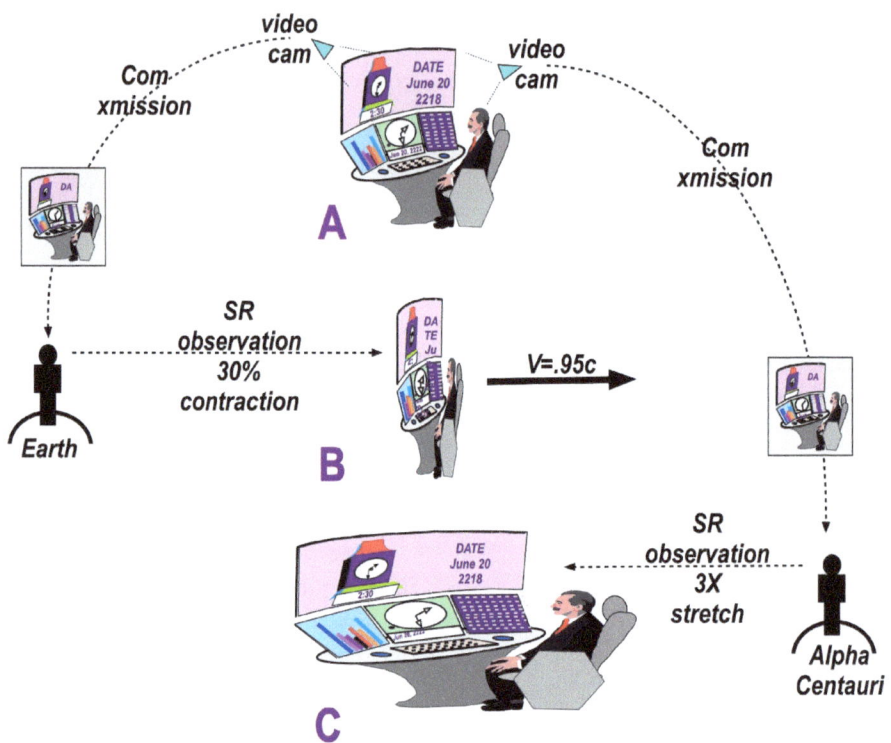

So, there are three supposed truths (**A, B** & **C**) that we are faced with. All of these observations play into the "twins' paradox". Note we have not addressed the aspects of the video cameras mounted inside the Starship and above the twin as shown in **Illustration 27**. Maybe, they can help solve the dilemma. The cams transmit pictures via the communication subsystem to both Earth and Alpha Centauri. When received by the Earth-bound twin and the person on Alpha Centauri Proxima-b. The video is completely normal. There is no evidence of contraction or extension. On second thought, if the Starship and its crew were indeed "squashed" or stretched, the mounted video cam dimensions would be changed as well. The result being that the transmitted pictures would offer no sure proof of which observation **A**, **B**, or **C** is true. It appears that the only way to determine the truth is to go on a high-speed round trip!

<u>All Aboard, ready for departure</u>
<u>Itinerary:</u>
Earth Departure: 2:30 PM, June 20, 2023
Proxima-b Arrival: 2:30 PM, Sept 6, 2027
Proxima-b Stay Duration: 1 year
Proxima-b Departure: 2:30 PM, Sept, 2028
Earth Arrival: 2:30 PM, Nov 11, 2032
Distance: 4 light years each leg
Cruise speed: 0.95c each leg

<u>Voyage Cruise Details</u>
Before we board the Starship, we are given the cruise details of the planned voyage and its communication operations with Earth and Proxima-b. Proxima-b is a small planet orbiting the binary stars of Alpha Centauri four light years (2.4E+13 miles) from Earth. **Illustration 28, "Round Trip Voyage Profile"** presents a simplified visual of the voyage.

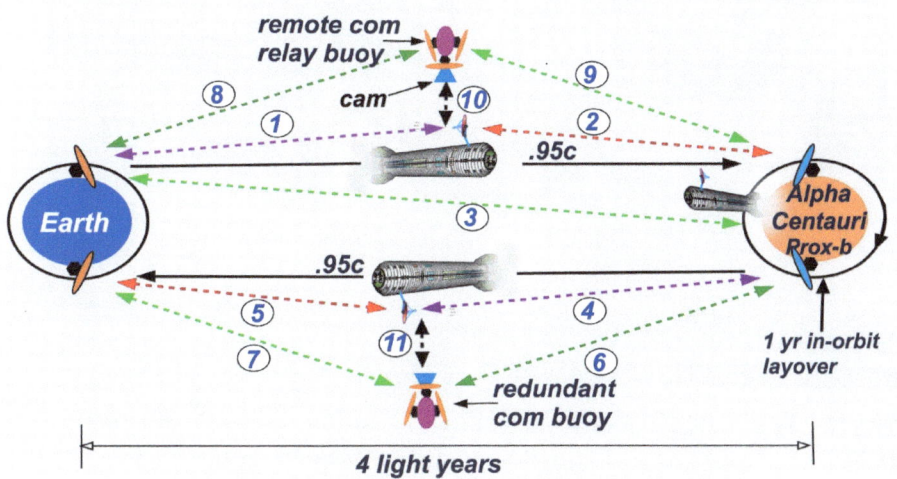

The first thing we notice is the number of communication links that are involved. These "com" links are numbered from 1 to **11** and are color coded violet, red, green and black. We also notice redundant "com" relay buoys (spacecraft) "anchored" about halfway. These buoys are unmanned and autonomous. They serve a dual purpose for backup links for the voyage and as relay signal power boosters. As we are about to board, we synchronize our watches to the Earth's town clock and the shipboard's clock. The twins also note that their heartbeat rates are identically sixty beats per minute.

Departure is exactly on time, 2:30 PM June 20, 2023. In less than a day, the Earth is far behind us and our sun is the brightest and biggest star in the sky. Later on, we walk back to the stern observation deck to view the quickly disappearing solar system. **Illustration 29, "A View from a Starship, 0.95c"** depicts looking aft at the receding scene. We are amazed by the view.

Illustration 29, "A View from a Starship, 0.95c"

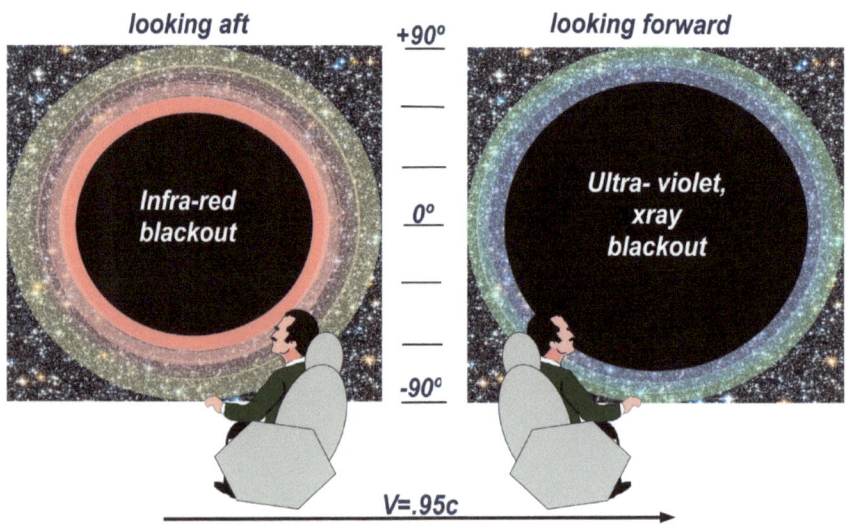

We are confronted by a "blackout" region where the solar system should be. No surrounding star field is visible within an angular region of ±50°. We go to the forward observation deck and we see a similar blackout region of ±70°. Again, we see a rainbow of colors due to Doppler effects of moving at a speed of **0.95c.** The question naturally arises; "How does the Starship communicate with Earth or Proxima-b under these conditions?".

Referring again to com links **1** and **4** in ***Illustration 28***, these communications are conducted with a "laser-com" system. Laser com links are used currently for some satellite links and they have been mostly in the infra-red range. For purposes of our discussion, we have assumed that both infra-red and violet laser-com links are fully operational and capable of communication over long distances. For these links, laser-com transmits and receives in the violet wavelength of about 400 nm (nanometers). Now for advancing situations, links **2** and **5** (color-coded red), transmit and receive in the infra-red wavelength of 800 nm. The black link between the Starship and the buoy utilize all three color links. It is important to note that the green color-coded links **3**, **6**, **7**, **8** and **9** do not require laser-com. They can utilize standard radio frequencies **(RF)** since the Doppler effects are small for these com links. However, for a Starship moving at high speed, laser-com links are a must to accommodate the large Doppler changes. However, a laser-com system would only be applicable up to speeds of **0.95c (for reception) to 1.5c (for transmission).** It would not be possible to communicate beyond these limiting speeds because of huge Doppler shifts! This is an ***unexpected*** development. Although this book has shown that there is no physical **SR** limiting speed c, it has shown that a completely new com system would have to be invented for traveling at the communication limiting speeds **(0.95c & 1.5c)** to maintain communications.

This author would like to pause and digress for a moment. There might be some promise to overcome this communication problem. It might be possible to use the *"entanglement"* phenomenon experimentally found in physics, Einstein referred to this physics topic as "spooky". Basically, entanglement entails the "destruction" of matter from a <u>unique single</u> event into subatomic particles. Some of these particles have been found to have mysterious and unbreakable bounds like twin brothers and sisters. The "spooky" aspect is that if one twin is destroyed, the remaining twin is destroyed as well. It gets even spookier. The twins could be separated by vast galactic distances, but if one ceases to exist the other *instantaneously* ceases to exist too. This implies that if the entangled pairs could be kept in separate containment systems, one containment system on Earth and the other on-board a Starship. By metering and destroying particles, maybe in a pulse coded manner, a com system can be established by detecting their entangled pair destruction. This revolutionary com system would be an instantaneous two-way link capable of operating over galactic distances. Just a thought, end of digression.

Below, we will address only three communication events between the Starship and Earth. Later we will discuss two more com events at the half-way points.

Com #1 At the Earth out-bound half-way point
Com #2 At entry into Proxima-b orbit
Com #3 At the Earth in-bound half-way point

CAUTION! 🚩 **Com #1:** *At the outbound half-way mark*
We are now approaching quickly the 2-year half-way mark on our out-bound cruise. The cruise has been uneventful up to this point and the decision has been made to continue. At exactly 2:30 PM shipboard time, a digital picture of the crew and Starship console clock will be transmitted to Earth and to Proxima-b. Likewise, a pre-departure arrangement was made to transmit a digital picture of the town clock to the Starship to arrive simultaneously at the half-way point. Remember, the actual cruise time to the half-way mark is 2.1 years at $V=0.95c$. Therefore, the transmission to Earth occurs at exactly the same time the com message is received from Earth. **Illustration 30, "Com #1 At Half-Way Outbound Mark"** shows the sequence of these message exchanges, Message (Com#1) was transmitted from Earth on July 28 2023 at 2:30PM and later received by the Starship on July 28 2025 at 2:30PM. At this same time, the Starship Transmits a message to Earth which is received at Earth 2 years later on July 28, 2027.

The messages include a normal video-cam digital snapshot and a digital telemetry readout of the twin's pulse rate. As is pictured in the exchange snapshots, everything is normal. Everyone is alive and well with a heartbeat of 60 beats per minute (bpm). **Special**

Relativity (SR) says that each person, whether on Earth or in the Starship moving at **0.95c**, cannot detect motion and normal physics apply. Hence the transmitted snapshots show the truth of the

Illustration 30, "Com #1 At Outbound Half-Way Mark"

situation. But **SR** dictates that something different is occurring depending on an observer's location. As shown in this Illustration, an Earth observer (a twin) "sees" the stated *mathematical* truth that the Starship has physically contracted by 70%. In other words, a 1000 ft long Starship has "shrunk" to 300ft. Further, the town clock seems stopped at July 28, 2023. In addition, any onboard clock (in this case a human heart beat) would have slowed down by a factor of three. Thus, the heart beat would be slowed to 20 bpm and the onboard twin would be aging much more slower (time dilation). **SR** also predicts that the twin on-board the Starship "sees" the reverse of his twin back on Earth. That is, the town clock slows and the Earth twin ages more slowly than himself. This is exactly the kind of dilemma that results from **SR** mathematics.

But let's complicate this dilemma even further. Let's consider the case of the human heart. As shown in **Illustration 30**, if indeed a person is "squashed" from a 1ft thickness (chest to back) together with his heart, what ramifications does this have. No physicist, as far as I know, has addressed this aspect. What happens when a person's thickness is compressed from 12 inches to 4 inches? Relativists would simply opinionate that **SR** is true and dismisses the question. But in actuality, no one really knows. You would naturally assume, aside from any pain, that a slow pulse of 20 bpm would produce a comatose—if not fatal— condition. If you think all of this is weird, wait until our next Com #2 event.

<u>CAUTION!</u> Com #2: *At entry into Proxima-b orbit*

We have finally arrived at our interstellar destination Proxima-b and established an orbit around the planet. Our second communication with Earth has just been operationally performed. **Illustration 31, "Com #2 At Entry into Proxima-b Orbit"** provides a picture of the message exchanges. Einstein and others never considered a stopover before returning to Earth. This planned stopover of one-year duration reveals other **SR**-related cautions. First, we notice that **Com #2** was transmitted from Earth 16 days after Starship departure on Sept. 6 2023 2:30PM. This date was again selected for simultaneous reception at the time of orbit entry. **Com #2** is received 4 years later on Sept. 6 2027 2:30PM. At this date, a **Com #2** message is transmitted to Earth with a future arrival on Sept. 6, 2031. Surprisingly, **Illustration 31** shows that every one is well on Earth and at Proxima-b. There are no comatose conditions and anyone that has passed away because of **SR** contraction are now resurrected. Note that there are no significant planet to planet Doppler effects present. But in looking at the snapshot of the town clock, we see it is showing a 4-year old image. And as each day goes by in Proxima-b orbit, the town clock remains 4 years in the past. There is **NO** lingering time dilation. The continuing time difference of 4 years is simply the four years it takes a light image to propagate from Earth to Proxima-b. There is no mystery and

Illustration 31, "Com #2 At Entry Into Proxima-b Orbit"

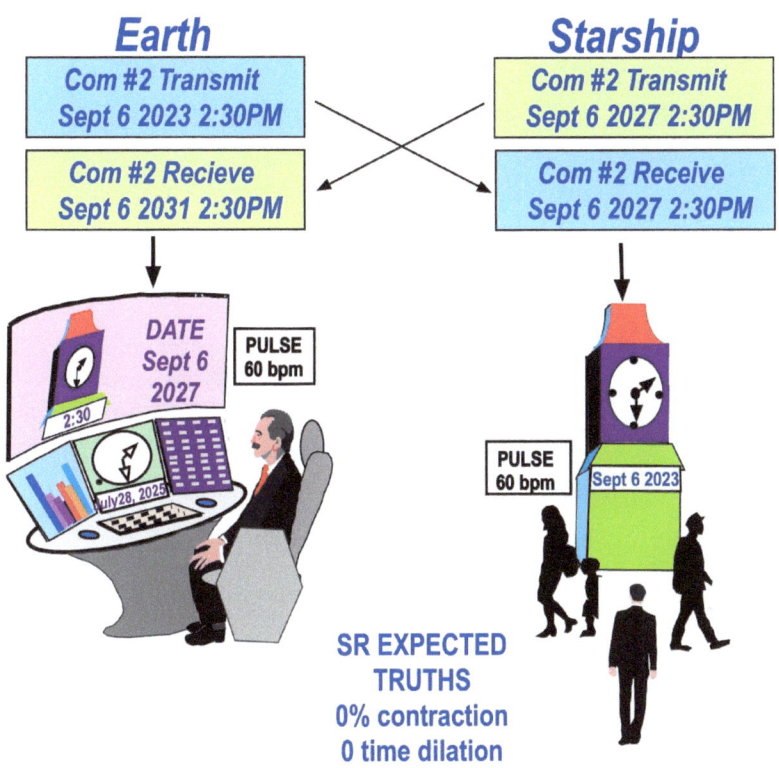

clocks remain synchronized. When **Com #2** from the Starship is received at Earth, they will see an image of the town clock displayed on the Starship console. They will question the implied facts that there was no slowing of onboard clocks during the outbound cruise. Again, I'm afraid that countless explanations that support **SR** will still be put forth. Most likely, there will be no reconciling of **SR** until an actual voyage at high speed will be accomplished. Now we will proceed to our next message exchange **Com #3**.

CAUTION! **Com #3:** *At the inbound half-way mark*
This com message is very similar to **Com #1** message at the outbound half-way mark. Therefore, it won't take much time to discuss. **Illustration 32, "Com #3 "At Half-Way Inbound Mark"** presents this event. It shows that the transmitted images appear normal. People are alive and undistorted. Pulses are beating at 60 bpm. However, when **SR** effects are applied, time onboard the Starship is speeding up. Our human heart clock is beating at a rate of 180 bpm, sufficient to cause extreme hypertension or stroke. So, our onboard twin is aging much faster than his twin on Earth. But everything is relative. Neither twin can tell if he is moving or not. The exact reverse is happening than **Com #1**.

Illustration 32, "Com #3 At Inbound Half-Way Mark"

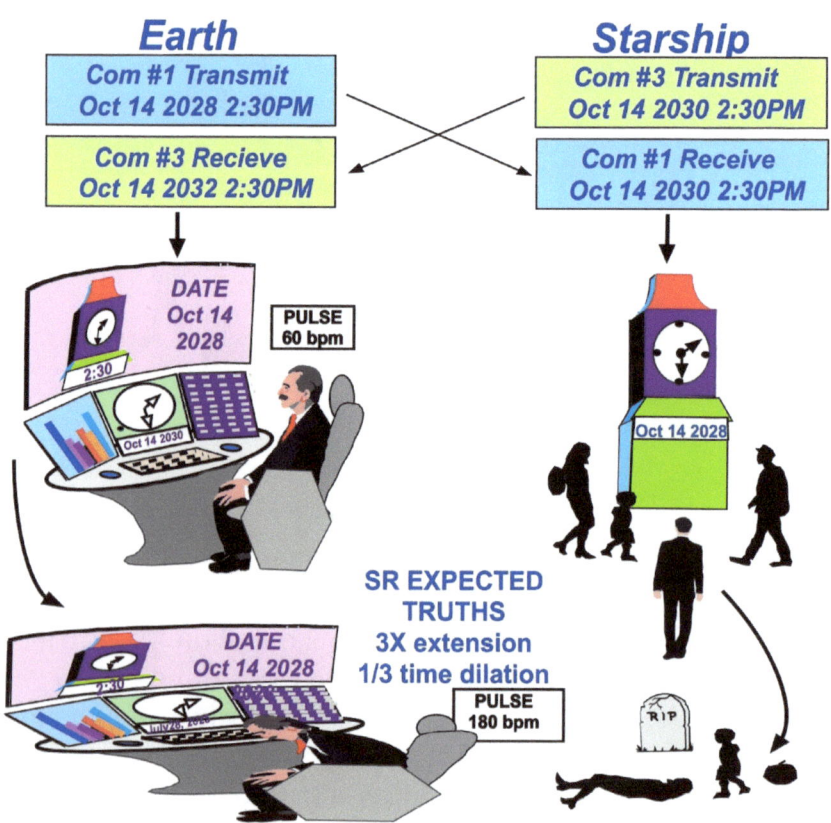

The world of **SR** is certainly disconcerting and non-intuitive. It is easy to become embroiled and confused with the back and forth of trying to keep things straight in one's mind. All of these Com exchanges are still insufficient proof that **SR** is not founded on solid ground. This brings us to our next com Alert to prove the matter.

ALERT! 🚩🚩 **Flaw #8 "Twin paradox debunked!"**
Certainly, to an interested party, the above discussion would be sufficient cause for concern of the validity of **SR**. But as mentioned above, "relativists" having been entrench a hundred years in **SR** and **GR**, would need proof-positive that there is an obvious error involved somewhere. And that is exactly what we are going to show! The whole question of any **SR** validation is whether or not physical lengths of objects shorten at high relative speeds. The attendant outcomes of this question include the validation or invalidation of the dilation of time and the increase of mass. All these correlated effects lead to the limiting mathematical singularity present in **SR**; i.e., no object can exceed the speed of light *c*. We have previously pointed out **SR** *Flaws #1, #2* and *#3* in **Chapter 3** that invalidate **SR**. But evidently, we have to "pile on" additional flaws to make a dent to seriously prove that **SR** theory is invalid. The twin paradox is the ideal "venue" to reveal these additional flaws with a proof-positive expose`. Hence, a double Alert Flag is shown to emphasize the importance of the following discussion and common-sense logic. As mentioned, the crux of the proof lies in the shortening or elongation of the Starship.

Illustration 33, "SR Contraction & Elongation, 0.95c", vividly conveys this **SR** length shortening effect in an unbelievable and ridiculous sight! It can be seen that at a speed ***V=0.95c*** receding from a fixed observer on Earth, the Starship is contracted to 30% (300ft) of its original design length of 1000ft. If the Starship is advancing toward the observer, its physical length would increase 3X to 3000ft. Again, note that the length changes are only in the direction flight, ***not in the width (diameter) of the Starship.*** This fact of the **SR** derived mathematics is the key to debunking the twin paradox and thereby, again, invalidating the entire **Theory of Special Relativity (SR)** and its mathematics. The only hope of proving **SR** and the twin paradox is to actually accomplish an interstellar voyage at high speed. Of course, this will not be possible for a hundred years or more. I again refer the reader to **Reference 1** quoted in <u>**Chapter 3**</u> "**Relativity!... Really??**". This reference book provides a rudimentary Starship design and round-trip interstellar mission. It shows the immensity, cost and technological advances needed to accomplish such a mission. In short, it would take a hundred years or more to wait for a definite validation of **SR**. Waiting this long is not acceptable! How can we prove or disprove the twin paradox now and avoid the wait time and trillions in cost?

Illustration 33, "SR Contraction & Elongation, 0.95c"

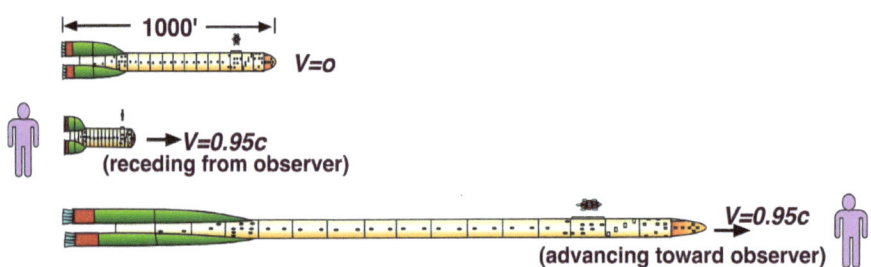

This author is glad to tell you the question can be answered today! The tact taken is not by validating with elegant and complicated physics and mathematics, but with a thought experiment. But unlike Einstein's thought experiments that led to **SR** and **GR**, this thought experiment uses common ordinary everyday proven technology. There are no postulations posed, and therefore mis-interpretations, refutations or ambiguities are completely avoided. The author's thought experiment is nothing more than applying space communications engineering to the problem at hand, namely resolving the twin paradox. That's why it was necessary to prepare the reader concerning communication principles and their application. We are now ready for **SR's** *coup de grâce.*

Now let's set the stage for our thought experiment. What was the flaw or culprit that led us to this paradox? It was Einstein's inability to understand what was happening when he was running along a light beam and encountered a "frozen" light wave (see <u>**Chapter 3**</u>). He didn't recognize that he was simply going through a "blink-out" phase of communication (see *Chapter 6*), and that he was dealing with Doppler effects. He then stated that this frozen condition was impossible and, therefore, put forth his famous Postulation 2 (see **Chapter 3**). Postulation 2, in the end, simply stated that all observers, whether moving or not, measure a light ray impingement velocity to be *c*. The consequences of the mathematics derived from Postulation 2 are the shortening of physical objects along the direction of flight *(V)* and the expanding of time to complete an observation. In other words, ignoring Doppler effects is the root cause of these consequences. Since there is no accompanied dimensional change in width (diameter), this perpendicular orientation is NOT impacted by **SR**! For this reason, our thought experiment approach is to minimize any Doppler effects in the conduct of the experiment, but yet be able to observe any **SR** related physical length changes related to *V*. How do we do this?

We set up our thought experiment by first revisiting the outbound half-way mark and take a closer look at the anchored com relay buoy. We see that the mission role of the buoy is to serve as an augmenting or backup link during the round-trip interstellar mission. Actually, it is a fairly sophisticated spacecraft able to operate autonomously for many years. It not only communicates with Earth, Proxima-b, and the Starship but with its redundant buoy partner anchored several million miles away. The buoys perform self navigation, station-keeping maneuvers to remain fixed in deep space. A great amount of Artificial Intelligence **(AI)** is employed to analyze and correct anomalies between itself and its redundant partner. As mentioned previously, the com links consist of radio frequency and laser com and LIDAR links. ***Illustration 34, "Twin Paradox Debunked!"*** depicts a single com relay buoy at its mission location. Com links to Earth or Proxima-b are not shown. In ***Illustration 34***, the Starship is navigated to pass between the buoys at a distance of about a million to two million miles. It is extremely important to note that the "fly-by" path will be perpendicular(90°) to the buoys at closest approach. Upon nearing and receding (within ±15° from vertical) at the closest approach point, a small window will open where the the Doppler effects of a moving source (Starship) and a fixed observer (buoy) or vice versa are minimized, well known, and easily compensated. Data processing is robust, uncomplicated and straightforward. It doesn't have to contend with unknown miniscule error sources. As

Illustration 34, "Twin Paradox Debunked!"

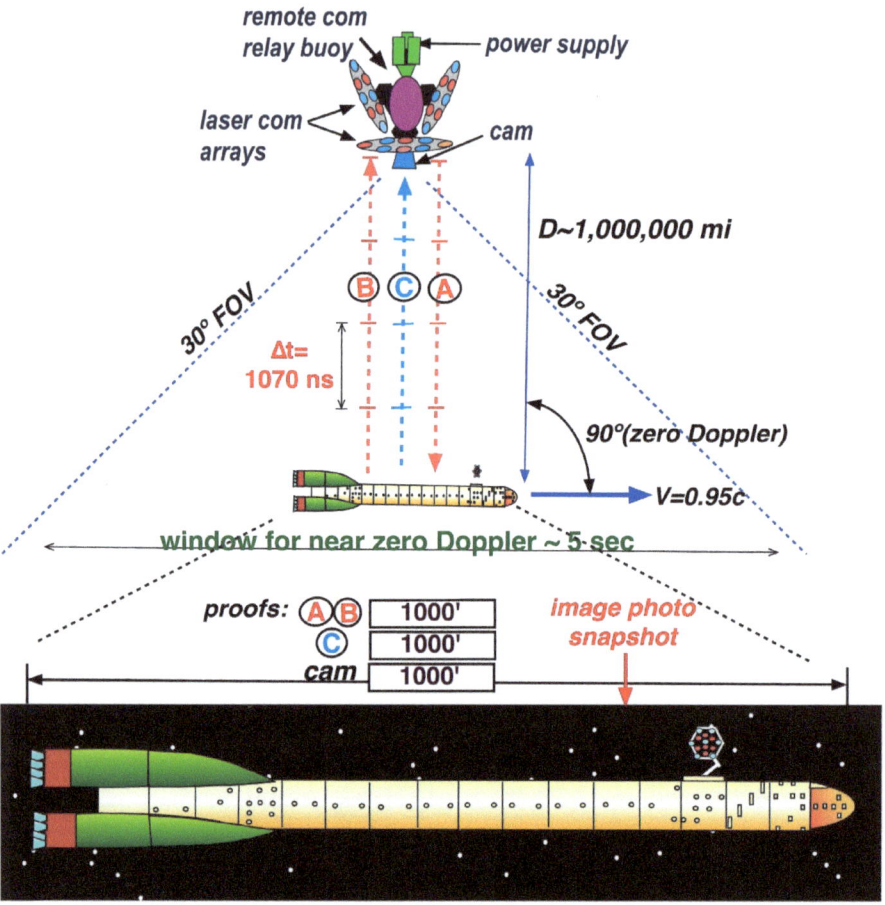

mentioned before, navigation, communication, and imaging technologies needed for the experiment are currently well understood although they face moderate engineering design challenges. The frontiers of technology that must be faced, lie in the areas of Starship Propulsion and human life factors. Thus. we have established the priority conditions for a successful experiment, i.e.; an **SR** parameter that is huge (1000 ft) by comparison to other "Einsteinian" parameters and robust straight forward processing. Now it remains to specify the realistic methods for measuring the length of the Starship.

Illustration 34 shows three measurement implementations. One is with a transponder transmit and receive link (**A** & **B**). However, this link is modified somewhat to provide the Starship length measurement. The modification entails a full Starship length "hard-wire" electrical signal path before the signal is retransmitted back to the buoy. So, if the Starship differs from its "at-rest" physical design, any real **SR** dimensional change will be measured proportionally with any actual shortening or elongation. Of course, since there are **no significant** error sources. Doppler effects and Starship range are easily determined, and therefore. the measurement can **only be a true 1000 ft** even if one uses Einstein's relativistic theory!

A second independent backup method is a similar one-way link (**C**) from the Starship to the buoy. It too has a real design signal hard-wire path length of 1000 feet. By hardwire, I mean a physical signal wire is connected 1000 feet from the front to the rear of the Starship. Again, if there is a physical Starship length change due to relativistic effects, the pulse frequency will also change. Now in both the transponder link and the independent Starship link, the time tagged signal pulse is every 1070 ns (nanoseconds). This value of 1070 ns is the time it takes light to propagate 1000 feet. Again, since the experiment is never more than ±15 ° from vertical, the true doppler corrected measurement will be 1000 feet! Now the author poses the following dilemma. If **SR** is indeed true, and the **SR** length measurement is greater than 300 ft, how do the two observers (Earth & buoy) reconcile the two *simultaneous* measurements? They both can't be true! Since only one can be true (buoy), the other (**SR**) must therefore be apparent!

Even this experiment scenario may still not be convincing to some staunch Relativists even if only true measurements results are being observed in the experiment. This brings us to the third measurement method of using a multi-spectral camera to determine Starship length. This method does not lie or produce a false-positive.

Referring to **Illustration 34**, note the camera mounted on the buoy. The camera system is not unlike a similar high resolution carried onboard the Hubble telescope spacecraft. This spacecraft was operational a decade or more at this writing. The resolution of the camera itself has been stated to be able to detect a pair of automobile headlights from five thousand miles away. Certainly, the camera to be utilized will be tailored to ensure obtaining useable "snapshots". In preparing for this imaging event, it would be very easy to ensure that the Starship would transit the camera field-of-view **(FOV)**. Perhaps, additional assurance could be added to the task by "painting" the Starship with an illumination beam from the buoy. Additional picture taking enhancements can include high - intensity flood lights or dispensed flares. As shown in ***Illustration 34***, a 5 second window would be available for imaging. This author believes that a snapshot could be taken within about 100 ns to insure resolution accuracy thus affording opportunity for multiple snapshots during the open window. A slewing or tracking function of the cam may also be possible to accumulate more dwell time on the Starship for imaging. The true conclusion that must be reached is that the length of the Starship is un-ambiguously 1000 feet as shown by the photo at the bottom of ***Illustration 34***. This same operation can be conducted on the Earth-bound trip as well.

All three methods of this thought experiment have confirmed the outcome that there is no change in the Starship length! The shortening or elongations derived by **SR** are simply apparent or imaginary. The flaws and fallacies of **SR** Postulations and logic have been expressed and proven incorrect repeatedly in **Chapter 3**, **Chapter 4**, **Chapter 6** and *Reference 1* given in **Chapter 3**. Since there are no Starship dimensional changes, it follows that there is **no time dilation** as well! Therefore, the **twins are the same age** when they meet again at the end of the voyage!

Paradox Debunked!
This discussion and the above chapters demonstrate how an errored assumption(s) can confound the interpretation of the mathematical and physics derivations. Further, any new or other theory that involves **SR** and **GR**, requires more confounding interpretations upon confounding interpretations. The result is that no human can contain the whole rationale and saneness because they would be mired in an entanglement of circuitous non-intuitive logic.

Summary

Congratulations, you've completed, what I hope was both an entertaining and rewarding journey. The prime intent of this book was to evaluate and explain the Einstein's **"Space-Time Continuum"** concept from an engineering perspective. The presentation style was designed to bridge the general interested public to the complicated and involved **Relativity** theories and paradoxes. To this end, complicated mathematics and physics phenomena were avoided and instead, common sense statements and extensive visual illustrations were used. Thus, this book's content is easily readable and understandable for ages from pre-teens to octogenarians. In the process of exploring the concept of **"Space-Time Continuum"**, certain non-intuitive discussions and illogical developments led to uncovering a major flaw. This flaw, in turn, prompted the relook at other **Relativity** subjects, such as; Einstein's **Special Relativity (SR)** and **General Relativity (GR)**, wherein other serious flaws and cautions were uncovered. The findings of this book can be summarized by the following visual *Illustration 35, "Space-Time Shattered"*.

Chapter 1 "Space-Time Continuum …What is it?" delved into this concept in depth. The discussion ranged from the origins of the concept to its implications that the whole universe is connected in a four-dimensional fabric of space and time and its relationship to gravitational mass. One of the conclusions reached by Einstein is that a man falling from a building is not pulled down by gravity, but rather, is pushed down by gravitational pressure from the **Space-Time Continuum**. This revolutionary idea raised eyebrows for sure, but currently is the accepted position of the mainline science community. However, **Chapter 1** uncovered a serious Illogical *(Flaw #1)*, in the concept. This flaw pointed out that the derivation of **Space-Time Continuum** utilized fluid dynamics theory around a *solid body* together with gravitational tensor stress mathematics. The consequence of this flaw is that gravitational forces cannot exist within a solid body like the Earth. Of course, in the **Continuum** concept, this cannot be true and must be considered to be errored in this regard! Note each identified flaw in **Chapters 1, 2, 3, 4, 6** and **7** is noted by a red Alert flag.

How did we arrive at such a shattering conclusion? Well, the **Space-Time Continuum** concept really rests on the foundational pillars of **Special** and **General Relativity (SR & GR)** as delineated In **Chapter 2** *"Run! The foundations are crumbling"*. These other involved factors led to weighing and evaluating not only **SR** and **GR** but to looking at actual published supporting experimental results and the famed **SR** "Twins" paradox. The results of relooking at these other factors surfaced and re-emphasized new and previously known flaws. A total of seven flaws were identified along with six serious experiment related cautions which led to overwhelming evidence that did not support or prove **Space-Time Continuum**, **SR** or **GR**.

Chapter 3 *"ALERT! Special Relativity flaws ahead"*, found three fatal flaws in **SR**. These flaws revealed that there is NO shortening of an object's length dimension and consequently NO stretching of time, i.e.; **NO time dilation**! The source of these flaws is Einstein's Postulation 2 which insists that physics operating in every frame is the same as a frame at rest in inertial space and that every observer *must* measure the impingement of a light ray as the speed of light *(c)*. This error, in turn, dictates that there are NO Doppler effects present when an observer is in a moving frame. But, computer simulation of light ray paths and their specific emitted photon quanta which are actually observed have proven that Postulation 2 is errored in its interpretation. Enforcing this postulation to having no Doppler effects on a moving observer deceptively results with the observer viewing an apparent path with the attendant length contraction, time dilation and the presence of a mathematical singularity in **SR**. This singularity says that no object can go faster than *c*, and that the mass of an object increases to infinity. These effects of **SR** make one squirm. How can mass be created? How can you compress an object without breaking it? What about effects on a human space traveler? If one atom reaches a near speed of *c*, does it become a black hole or wipe out an entire galaxy? What happens when the space traveler comes to rest again at his destination? Does he again become elongated? Do physical clocks speed up again, etc.? However, one must answer this common-sense question; *"How does a light ray that has no intrinsic connection to an object exert any physical influence within objects (their internal physical makeup) including clocks?"* A second follow up question is, "If there are no observers present, say to witness an speeding asteroid, is **SR** still viable?" Actually, by having proved **SR** is not valid, all these questions go away and this weird world of **SR** is returned to normal.

Now, moving on to **Chapter 4**, *"ALERT! General Relativity flaws ahead"*. **Chapter 4** demonstrated that Einstein's "Equivalence Principle *(a=g)*" is also fatally flawed! This principle is the basis of **General Relativity (GR)** which also is a supporting pillar for the **Continuum** concept. Essentially, the "moving elevator" paradox is shown to be in error. One elevator is stationary and at rest *(1g)* on the Earth's surface. The second elevator is accelerating*(a)* upward at **1g** in inertial space. It was proven that trajectories of dropped balls inside the two elevators are NOT the same! The proof lies in the subtle difference in the following statement.

"In the at-rest elevator, the ball is hitting the floor; but in the accelerating elevator, the floor is hitting the ball."

In the case of the accelerating elevator, the rebound force is twice that of the rebound force of the at rest elevator. It turns out that the ball's trajectory in the at-rest elevator is under the acceleration of gravity *(g)*, but when the the ball in an upward accelerating is released, it is in a free fall trajectory. Although the trajectories appear to be identical to either man, they are only deceptively identical. This conclusion is also applicable to balls "lobbed" horizontally across the elevators as well. Therefore, the **Equivalence Principle** is NOT valid and should not be used in any way as an underpinning to **GR** or to the **Space-Time Continuum** concept!

Now the reader might say, "Why haven't we heard about these flaws before now?". The answer is that **SR**, **GR** and **Space-Time** are marketable products. Believe it or not, there are many current detractors to **Relativity**. There are numbers of notable scientists, physicists, mathematicians, engineers and organizations who have taken opposing views with supporting research articles and published papers. ***Chapter 5***, ***"Oh, the error of our ways"*** gives a detail discussion concerning the two groups comprised of **Relativists** and anti-**Relativists**. The reader is strongly urged to read **Chapter 5**'s message about these two groups and various past conducted experiments. It will help immensely to understand the purpose of this book and the reviewing of related experiments that purport to prove the theories of **Relativity**.

Chapter 5 uncovers serious concerns about the conduct or the outcomes of certain experiments. Because the proof of Einstein's theories involved extremely small **Relativistic** parameters they are, unfortunately, often buried in the mire of experiment error sources many of which are greater than the error source itself. Hence. the title of this Chapter. Since no experiment, thus far, is robust and obvious enough for a non-interpretive, error source limited, non-contestable outcome, proof-positive validation remains illusive. It would not surprise this author if such an experiment must wait for an actual implementation, e.g. a high speed manned round-trip interstellar mission to prove that moving clocks slow down. For reasons cited in **Chapter 5**, three experiments were reviewed and assessed with what we know currently. These three experiments are listed below. Einstein himself identified these experiments and stated that if any one failed to produce a **Relativity** parameter confirmation, then **all of Relativity is failed.** The fourth experiment (Gravity Probe-B) is included as a recent "modern" satellite mission to prove the existence of space curvature and frame dragging of the **Space-Time Continuum**

1 Advance of the perihelion of Mercury
2 Deflection of light by a gravitational field
3 Gravitational red shift
4. Gravity Probe-B (GP-B)

As mentioned above, six serious concerns affecting the complete acceptance of poof for each experiment listed are raised. Because no experiment can be redone or revaluated, each uncovered concern is marked by a yellow caution flag.

Since the discussions of these experiments are technically detailed, the reader is directed to the individual experiment particulars in **Chapter 5** for a fuller explanation. Yellow caution flags are also displayed in **Chapter 7** as well.

Perhaps, the most perplexing paradox is the **"twin"** paradox. In this paradox, one twin travels on a high-speed space voyage. When he returns, he finds that his twin (who remained on Earth) is much older than himself. This paradox embodies most of the non-intuitive, illogical and noncommon-sense imaginations that are contained and required by **Relativity** theory. **Chapter 6**, "*The twin paradox, What is it?*", traces back the origin of this paradox, and identifies *Flaw #7* which proves that a "frozen" light wave is simply a "*fade-out*" phenomenon of the light ray and not a speed of light *(c)* limitation of any kind. In fact, any such limit on *c* is not a mathematical, imaginary or a physical one, but a practical one of maintaining communication links within $V_{max}=1.5c$ as explained on **Chapter 7**.

Chapter 7, "The twin paradox debunked", first provides a tutorial for the reader regarding space flight communications. This tutorial is of paramount importance in understanding the truths of a manned interstellar round-trip voyage. No postulations need to be stated to debunk the twin paradox as are present in **Relativity**. Every operation and physical conduct of the round trip are based on actual operations that are occurring billions of times every day in real life. The reader is taken on a near virtual illustrative voyage with celestial vistas as he speeds through space at ***0.95c***. Laser com communication links are used to send messages and images between the Starship and Earth. The confusion caused by communication delays over long distances of several light years are eliminated. Three communication locations are selected during the voyage to examine and demonstrate the communication link operations. The first location is at the outbound halfway mark and the second is at departure from Proxima-b. The third is at the return inbound halfway mark. Digital images of the Starship control console and the Earth's town clock at each of these locations are tracked from when they are transmitted to when they are received. Thus, the reader is given a true picture of what is happening as if it was occurring in real time. Even with this true portrayal and caution flags along the way of the voyage, it is still insufficient to convincingly validate or invalidate **SR**'s twin paradox. How do we obtain indisputable evidence to debunk the paradox?

The undeniable solution is to place an observer standing at rest outside of the moving Starship. By doing this, the **SR** peculiarity of length contraction can actually be shown to validate or invalidate itself. This is exactly what **Chapter 7** accomplishes. Prior to the manned interstellar voyage, redundant communication buoys were anchored at the halfway locations on a previous mission. The Starship navigates between the buoys and passes between them. Near the point of closest approach, the Doppler effects on light ray paths or laser com links are minimized and can be handled within the link capabilities. **Eureka!** An experiment is now devised wherein any Doppler effects of a Starship moving at near light speed ***c*** are completely known and accounted. During a small window of opportunity, a time tagged laser com signal equal to the Starship's design length is transmitted from the buoy is received and retransmitted by a transponder onboard the Starship. At the same time, the Starship transmits a separate one-way signal to the buoy at a pulsed frequency equal to the Starship design length. In addition, an image is taken by the buoy with a high-resolution camera. **SR** dictates that the length of the Starship must contract to one third its design length along the direction of flight. But since the direction of flight at closest approach to the buoy is perpendicular to the buoy's line of sight, all three methods of measuring the Starship's length from the buoy are identical to its design length. *Actually, there is no need to fly this mission!* There is nothing to prove! This thought experiment of **Chapter 7** is sound from both a scientific and engineering perspective. No assumptions or questionable postulations have to be pronounced! There is NO contraction or shortening of length and consequently there is NO time dilation! ......**QED!**

I hope you have enjoyed your travel through this book and that it was an interesting journey. Now, for this author, it is time to return to the reality of life, like minding a 55 mph speed limit.

The End

www.ingramcontent.com/pod-product-compliance
Lightning Source LLC
Chambersburg PA
CBHW040221220526
45473CB00001B/78